1

Corpus Venn 2

Contact Information:
diogodesouza7@gmail.com
diogodesouza7@hotmail.com

Table of Contents:

Introduction	2
Device	5
Philosophy	21
Electron Adventure	42
Numbers	49
Differential Equations	66
Zyn Tablets of Inquiry	84
Problems in Mechanics and Relativity	154
Revolution of World Progress	189
Quantum and Oscillations	220

Introduction:

We live in a world that is filled with technology and a high level of Scientific Understanding that has changed the world completely in the last 300 years since the European Renaissance and Imperialism. Telecommunications which are made possible through Transistors inside Electronic Devices and Electromagnetic Radiation propagating through space from Antennas, has connected all parts of the world facilitating communication like never before in human history as seen in the World Wide Web. The amount of complexity in these technologies are striking, and brings to mind that nothing is truly impossible for Science and Human Endeavor that may even perform Space Travel in the near or far future. When seeing the operation of a car's engine or the motherboard of a computer it is evident how much has Science allowed human control of Natural Laws. Now it is possible to control the flow of gases, liquids, or charges, and by taking advantage of the Natural Laws that drive the cosmos, we have machines doing the work for us fulfilling tasks may it be for transportation or telecommunication.

The number of Transistors in a Microprocessor are in the order of several Billions, more than a human can ever count in his or her lifetime. These Transistors work inside Chips similar to how Nerve Cells such as a Brain's Neurons work. It is this miniscule transfer of information in binary code of a series of 0s and 1s that when added to thousands, and millions, and billions of these signals allows for a greater Data Storage in our Electronic Devices, and transfer of great amounts of data that allows the function of Software. These signals send commands through pathways of a Computer's Hardware using a Language of Logic Steps, On and Off Switches from Transistors, that drive the thinking process of machines. With enough Transistors and a well elaborate set of commands, it is possible to have these machines learn things by themselves and become a being on their own which is the concept of Artificial Intelligence. The belief that our human brain is also a set of commands from the molecular transfer and sparks of energy in Neurons, a computer also operates through a series of mechanical steps. Common sense then revels that with enough Transistors and a greater complexity of 0s and 1s, Machine Learning is made possible and also the creation of a Machine that has its own thoughts and that can have its own personality and that can make its own decisions. That is called a living robot. Computers are evidence that nothing can stop human evolution, and even though space is enormous and space travel is currently extremely difficult, a faith that there should be a set of laws that can be helpful for us to be able to travel these long distances is what will permit human exploration to endeavor beyond the Moon into other Planets. In the next few pages, I tell a story of Particles travelling through a Device that gives a brief explanation on how awesome is technology now days, and how Philosophy can reach the realm of Electronics to provide meaning to our everyday life and our concept of reality. The device in the story exists only in this fiction with the sole purpose to bring admiration to the infinite possibilities or our Scientific Pursuit.

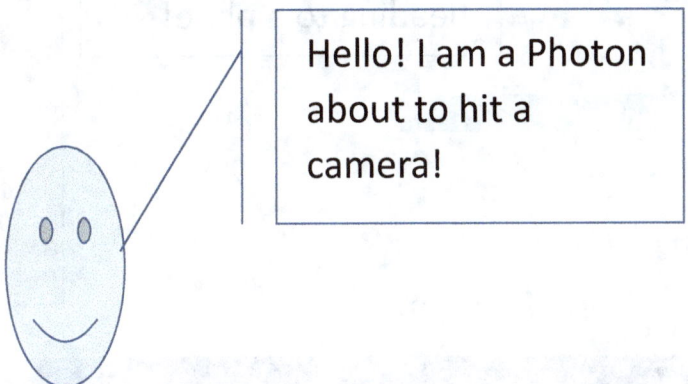

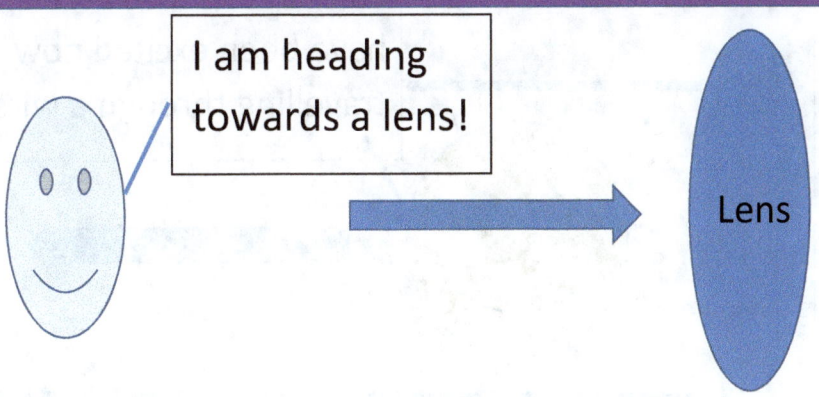

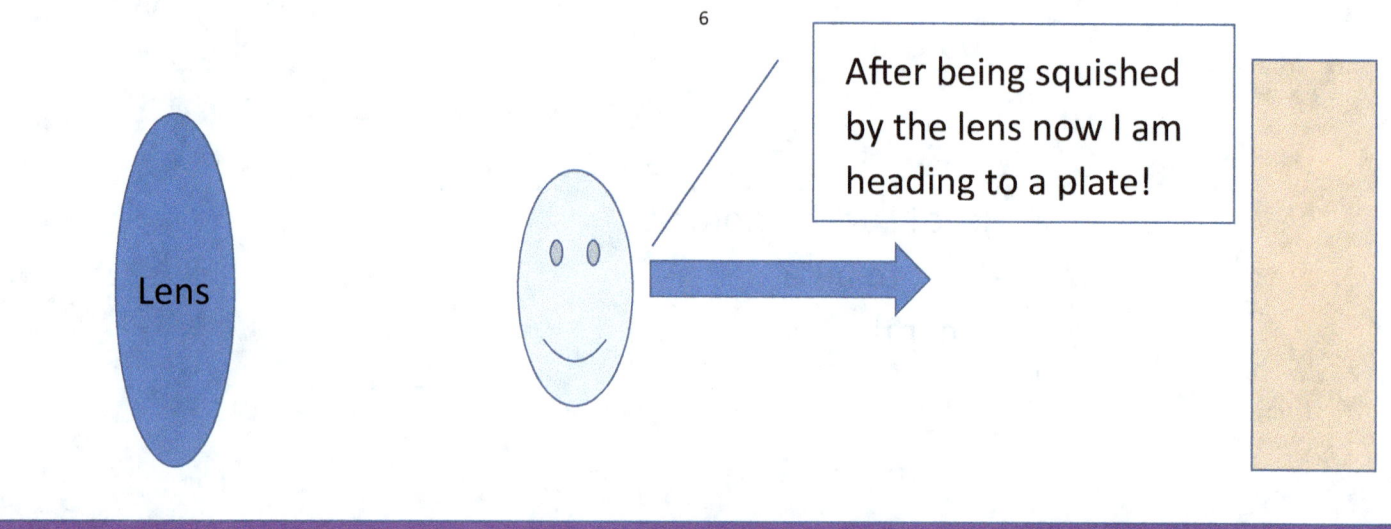

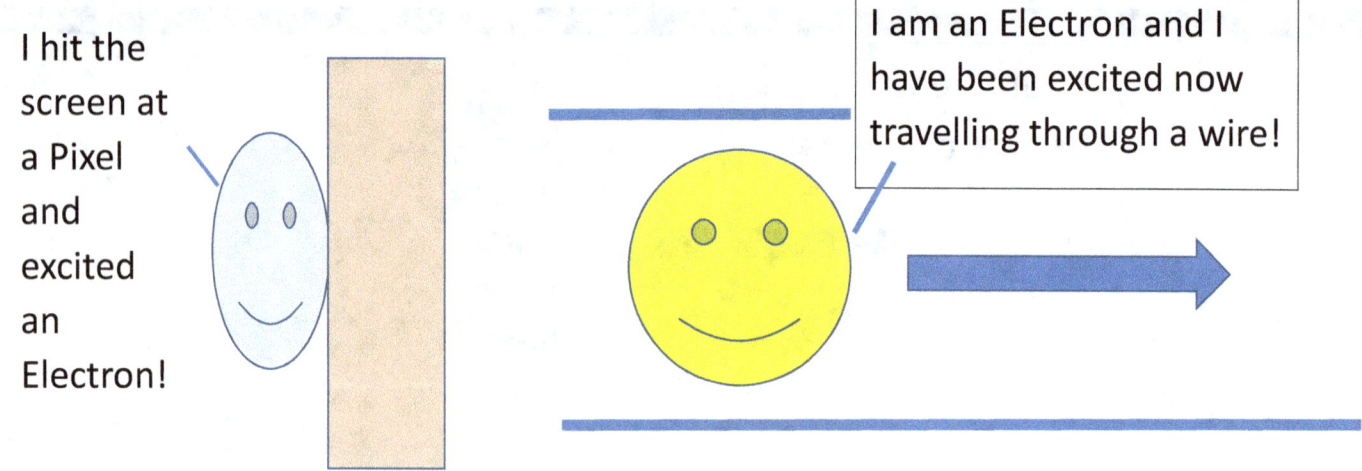

At the end of the wire, I entered a coil surrounding a metal and my presence there created a Magnetic Pattern in a tiny portion of a Hard Disc that is spinning in the direction of the blue arrow. I gave that portion a 1 value with an induced Magnetic Field towards the left as shown in the orange arrow.

Tip of the Needle that reads and writes on the Hard Drive Disc

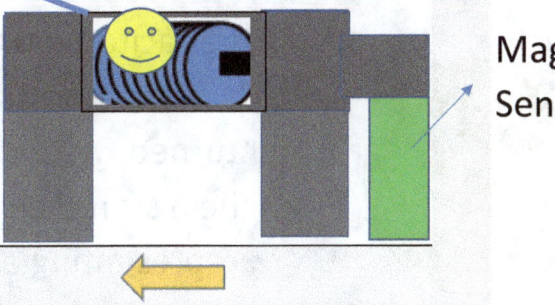

Magnetic Sensor

Disc

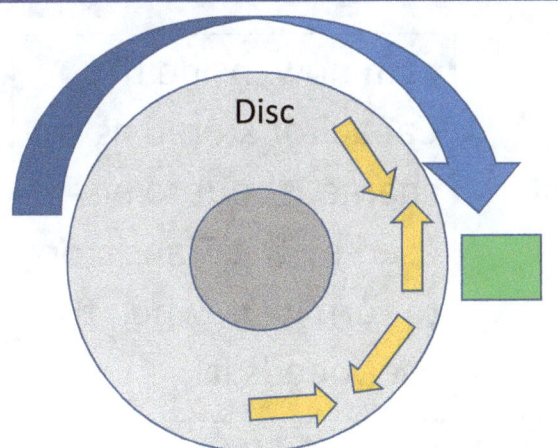

Hi! I am another Electron that was sent as a pulse by the green Magnetic Sensor from its change in Resistance that detected the Magnetic Field in the orange arrow from the rotating disc!

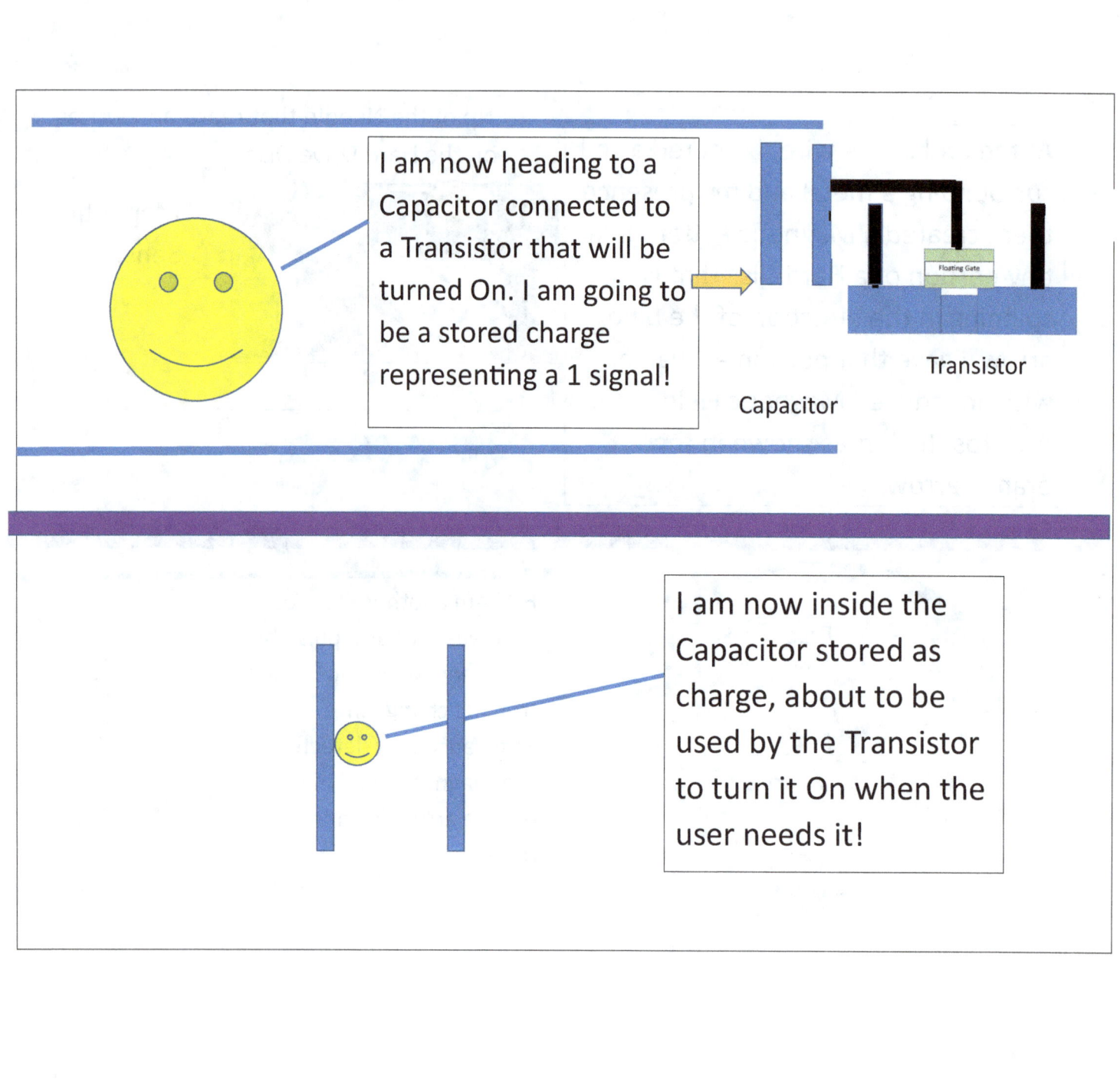

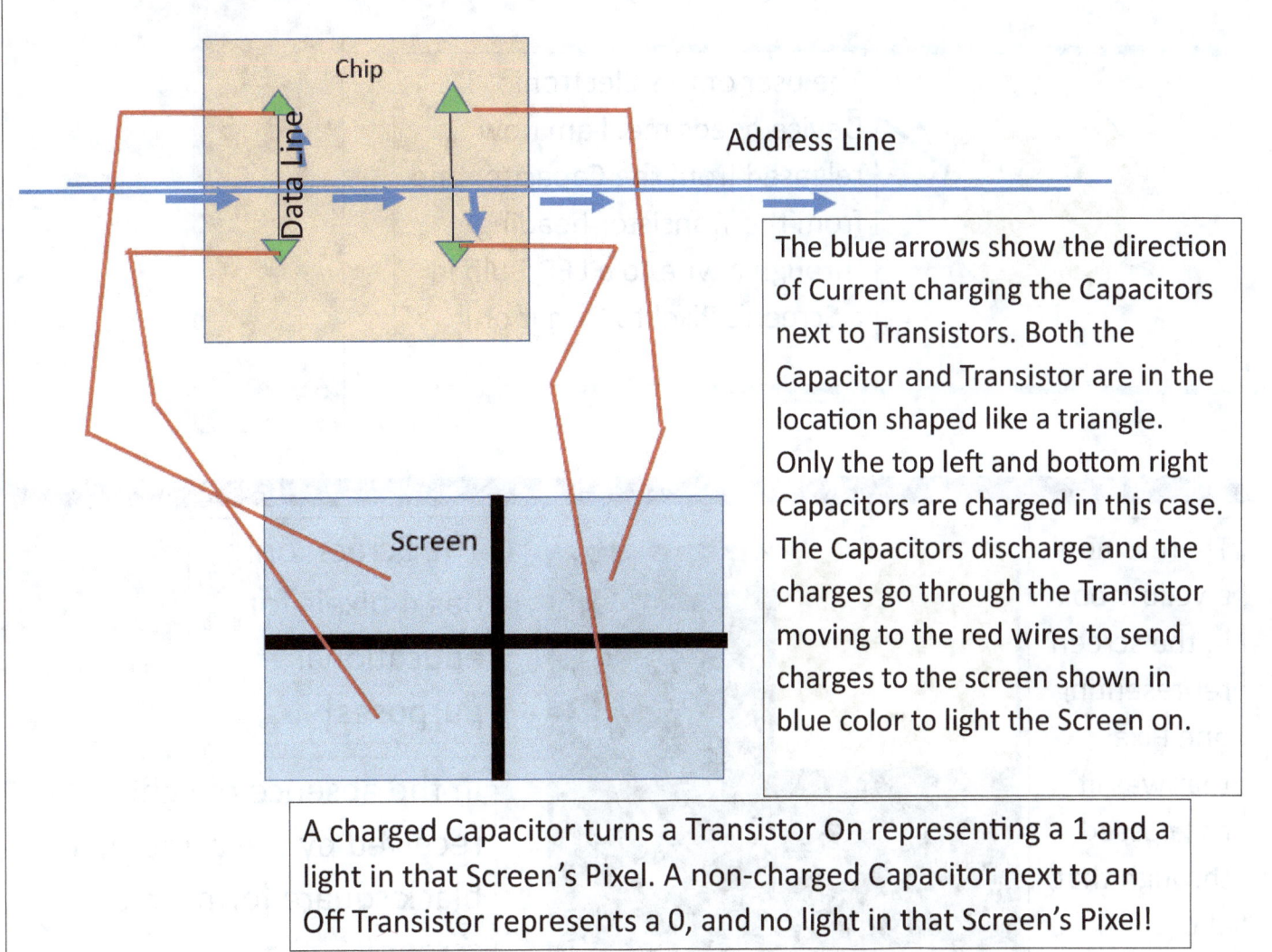

Chip

Data Line

Address Line

The blue arrows show the direction of Current charging the Capacitors next to Transistors. Both the Capacitor and Transistor are in the location shaped like a triangle. Only the top left and bottom right Capacitors are charged in this case. The Capacitors discharge and the charges go through the Transistor moving to the red wires to send charges to the screen shown in blue color to light the Screen on.

Screen

A charged Capacitor turns a Transistor On representing a 1 and a light in that Screen's Pixel. A non-charged Capacitor next to an Off Transistor represents a 0, and no light in that Screen's Pixel!

The user of this Electronic Device needs me. I am now released from the Capacitor and from the Transistor heading through a wire to a LED Bulb in a Screen's Pixel to turn it on!

There is now a yellow box in the screen representing one Pixel that was lit as I passed through that LED Bulb!

The Screen only has 4 pixels for educational purposes!

In the absence of light received by the camera, a black square forms in the screen!

Here in this screen with only 4 Pixels there are two black squares representing missing light, and two yellow squares representing the presence of light!

As the Hard Drive Disc spins it saves the Magnetic Orientation for the other Pixels in the screen. The information is then released generating a complete image!

Each Pixel is connected to a Transistor that sends the stored charge from Capacitors to the Pixel that flashes on in the screen!

Summary of the First Story:

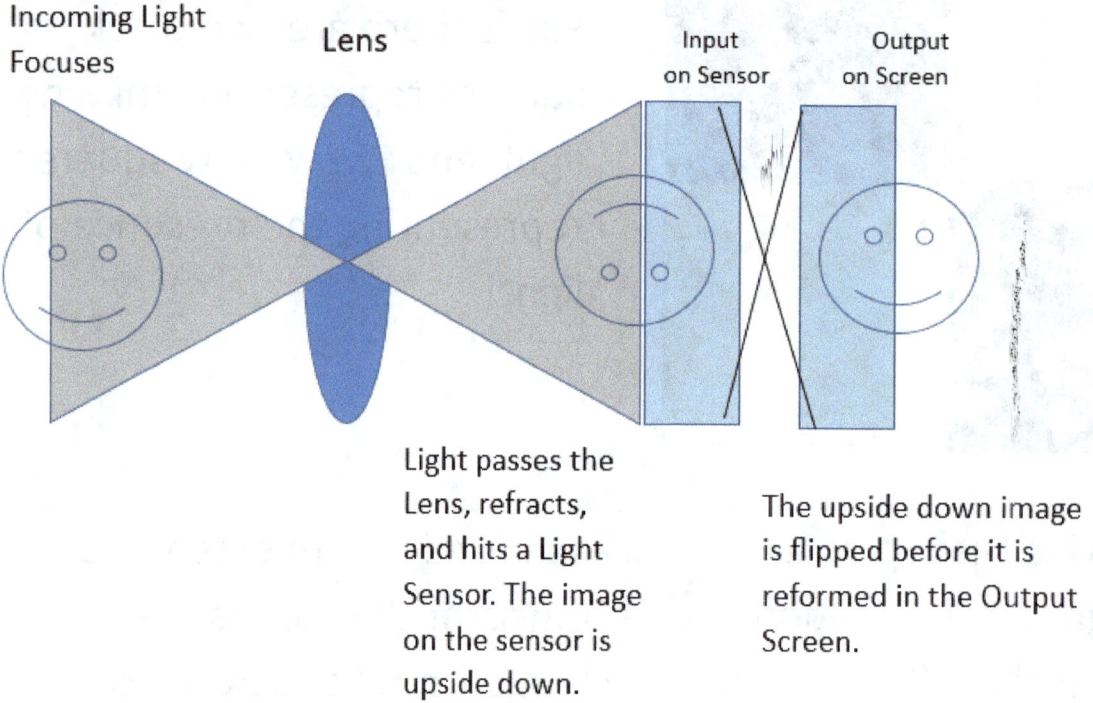

Incoming Light Focuses

Lens

Input on Sensor

Output on Screen

Light passes the Lens, refracts, and hits a Light Sensor. The image on the sensor is upside down.

The upside down image is flipped before it is reformed in the Output Screen.

Light travels towards the camera and is focused by the lens. The light refracts through the lens and diverges towards a CCD Sensor Plate. The information is gathered by the device that is released as output in the Screen.

CCD

Input Image Sensor

Output Screen

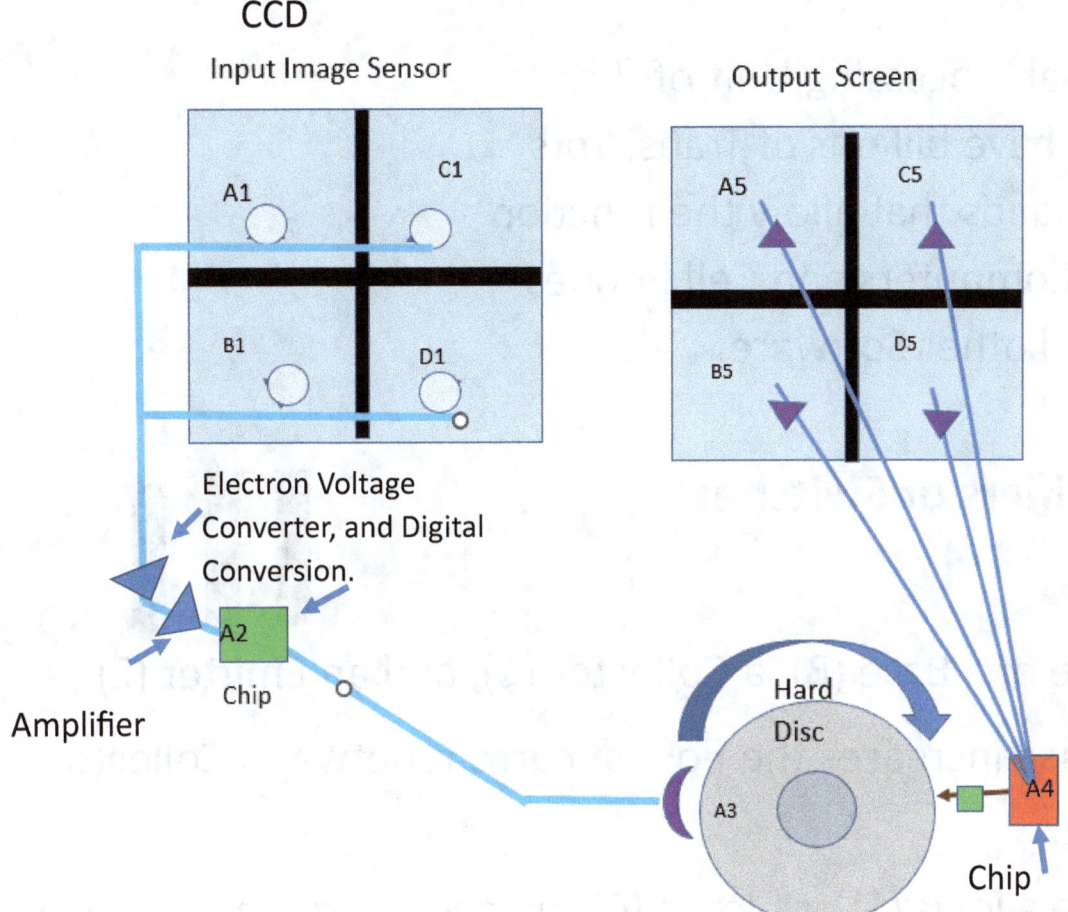

Electron Voltage
Converter, and Digital
Conversion.

Amplifier

A Photon hits a Pixel in the CCD Image Sensor. There are 4 Pixels shown. This Light is converted to a Current, passing through an Amplifier, and Converted to a Digital Signal of 1s and 0s. The information can be saved on a Hard Drive Disc, and later used to transfer the Digital Signals from a Chip to an Output Screen illuminating a LED Bulb for each Pixel.

What are CPUs?

CPUs are the Central Processing Unit of the
Hardware that can have billions of Transistors
with a set of commands that allow the function
and processing of Computers and Cell Phones
in running APPs and other Software.

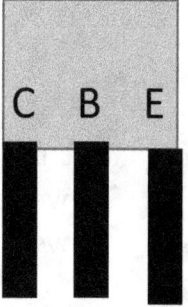

Transistors: Amplifiers or Switches

A Transistor is made of a Base (B), a Collector (C), and an Emitter (E).

In Amplifiers the Base increases the flow of current between Collector
and Emitter.

In a Switch the Base allows (1) or blocks (0) the current to flow between
Collector and Emitter.

Parts of a Transistor: EBC

A Transistor has three terminals called the Emitter, the Base, and the Collector. It is the Base of the Transistor that regulates when the Transistor is On or Off, and how much will the Signal be Amplified or reduced.

Types of Materials: Conductors, Insulators, and Semiconductors

1…**Conductors** which allows Charges to flow freely.

2…**Insulators** which does not allow Charges to flow easily.

3…**Semiconductors** which facilitate Charges to flow but not as much as Conductors.

Reading a CD and transmitting Sound

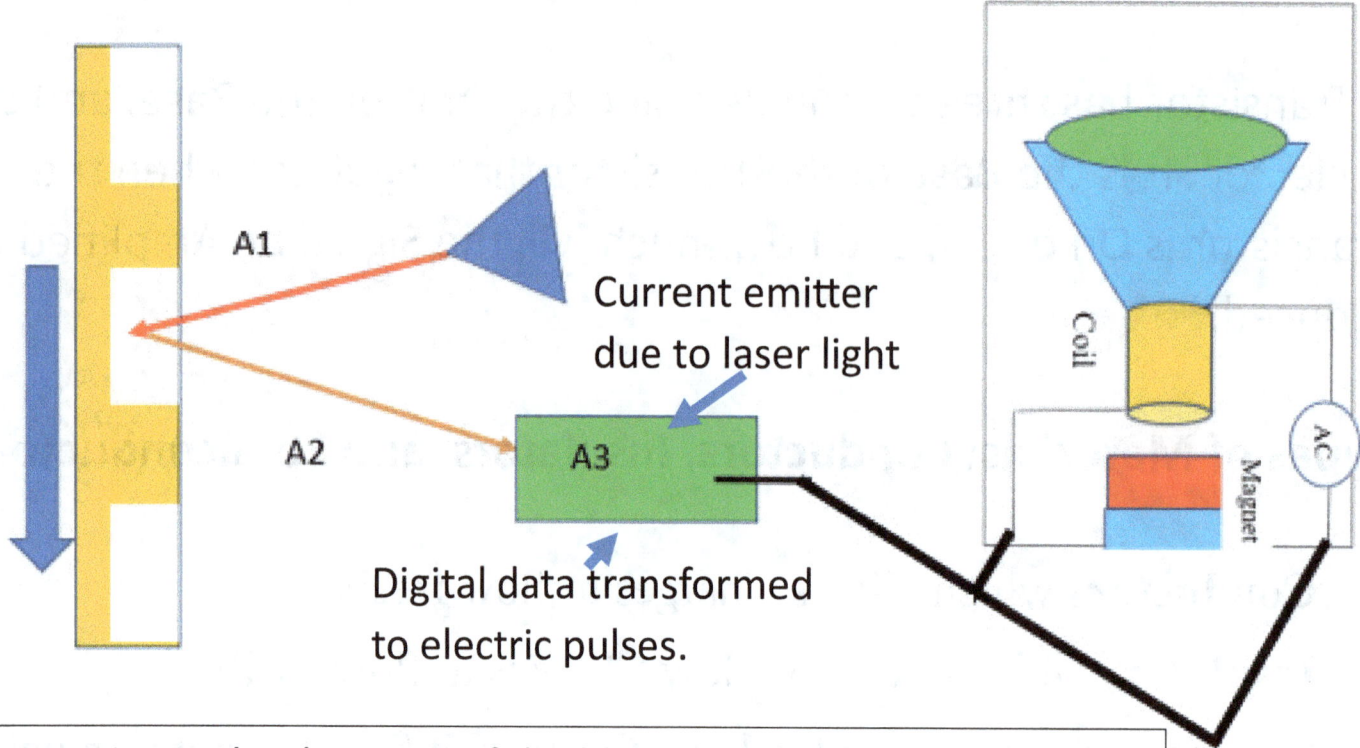

A1

Current emitter
due to laser light

A2   A3

Digital data transformed
to electric pulses.

A CD spins in the direction of the blue arrow. A laser hits the CD and when the beam hits cliffs it interferes constructively with the reflect beam, and when in the valleys interferes destructively. These patterns generate alternating current that is transferred to a speaker as Sound Waves.

Speaker

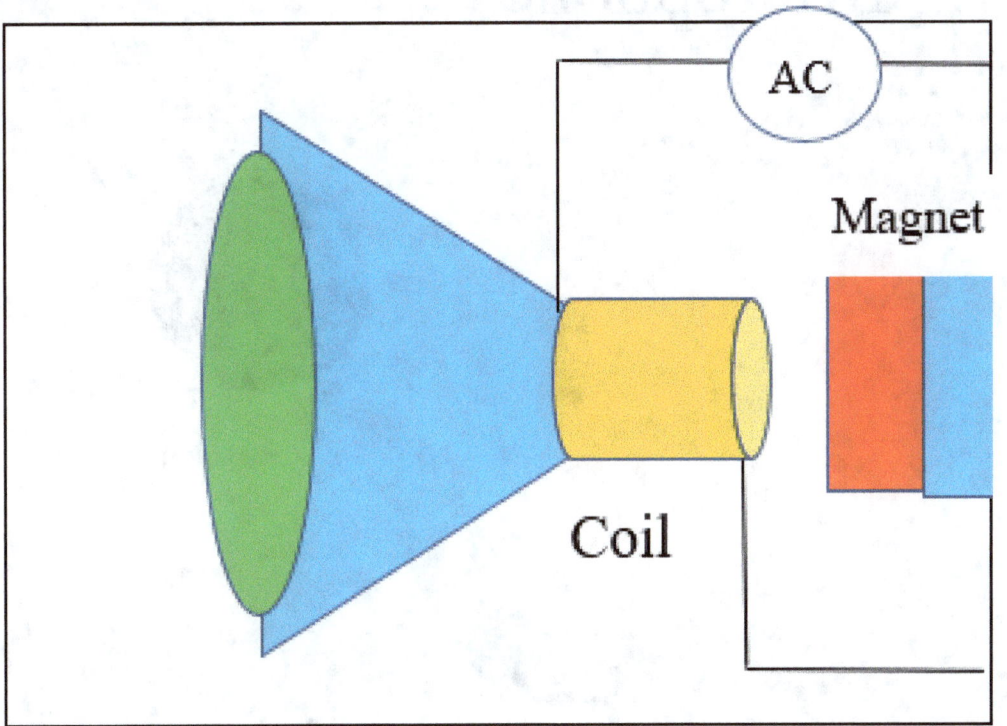

A Speaker converts an Alternating Current into Sound by forcing the Coil up and down against and towards the Magnet. This up and down motion vibrates a membrane that vibrates the particles in air leading to Sound.

Microphone

The Microphone works in reverse. Sound forces a membrane up and down, which moves the Coil above a Magnetic Field up and down, leading to an Alternating Current converting Sound into Electricity.

Transistor: **The Base is the Gate of Transistors**

ON: **Gates can be open or closed**

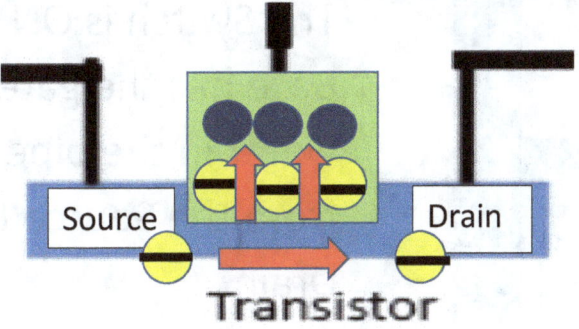

When Positive Holes arrive at the Base, they attract Electrons. The Electrons move up leaving a gap behind that is filled with the flow of Electrons from the Source to the Drain and the Transistor is ON.

OFF:

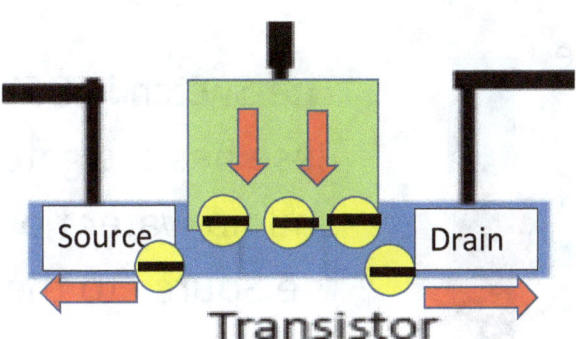

When Electrons move down from the base they repel the Electrons in the Source and Drain and no current can flow. The Transistor is OFF.

On and off switches: Suppose a Transistor is like a Pipe

ON:

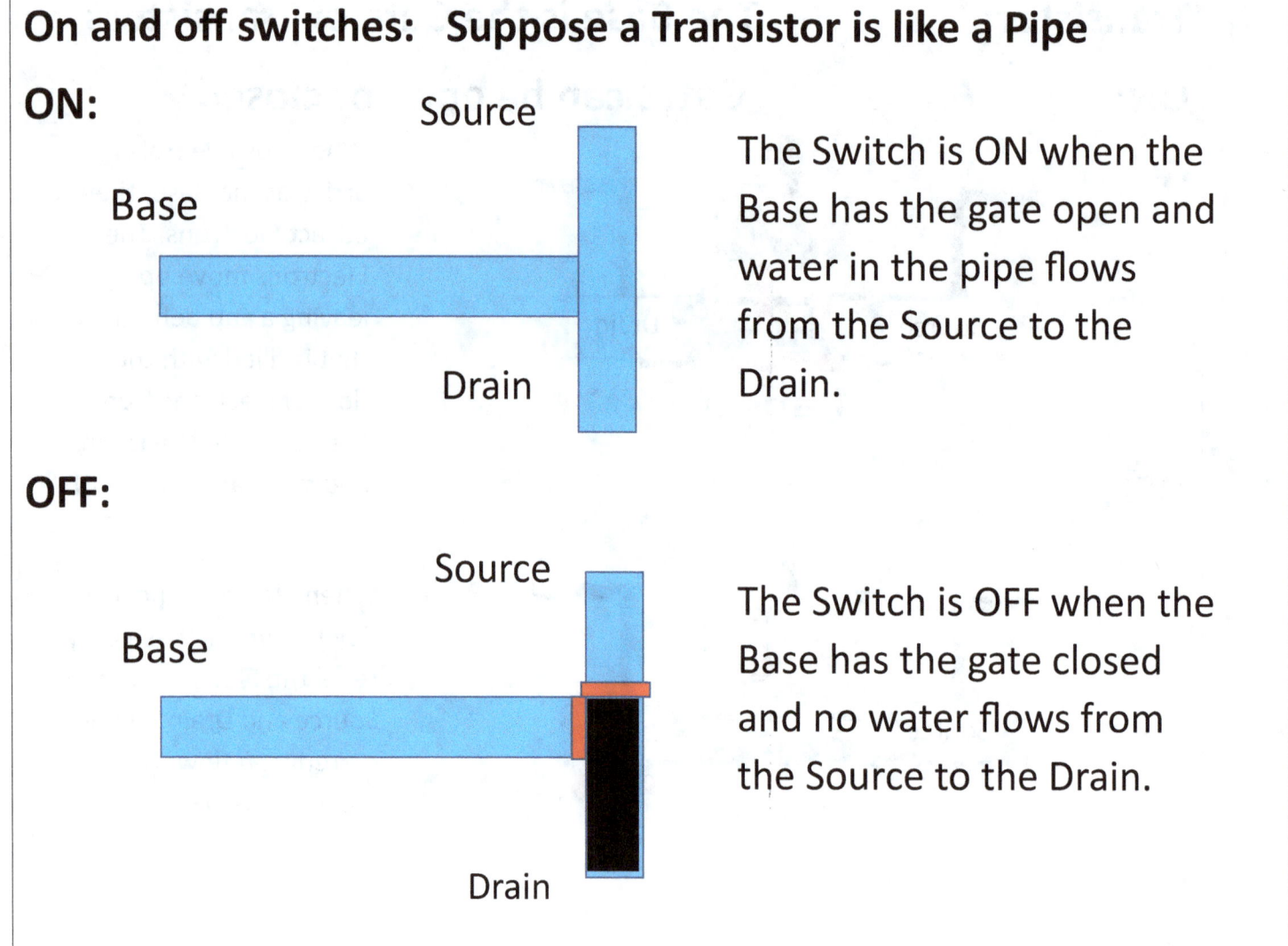

The Switch is ON when the Base has the gate open and water in the pipe flows from the Source to the Drain.

OFF:

The Switch is OFF when the Base has the gate closed and no water flows from the Source to the Drain.

Philosophy of Electronic Machines:

Today's World is filled with Technology where now it has become impossible to apply for a job, work, or schedule an appointment without the use of some APP while using an Electronic Device. When going to work we may ride in an Automobile, and when travelling overseas we ride on Airplanes or Ships. There simply is no way to survive in this New World without the use of these machines that work through Physical Principles in discovered Laws that rule the motion and behavior of Particles in the Universe. It is at no surprise that Modern Day Philosophers would be expected to generate many ideas and Philosophies focused on these newly discovered Principles. Most of Philosophy today, however, is focused on the exact same topics of centuries ago, and rarely a Philosopher delves deep into the Science of these machines. As a reader of worldwide ideas, I confess that I am tired of reading metaphysics, and Philosophy of Religion, and I feel the need of more Philosophers that will discuss the entire implications of the Nature of Science and Physics as seen in Electronic Devices. I refer to a deep inquiry about the whole process in which a Microchip works, or how electricity is generated, or how a Car's Engine work. This part of Philosophical Inquiry is very lacking today. The Ancient Greek Philosophers such as Plato and Aristotle did on my opinion, a great job in trying to understand reality and the natural world with Philosophy which at that time was linked to Science. Today's Philosophy, however, is rarely in the realms of Science, and when it does mention Physical Laws and Principles, it rarely delves as deep into the structure of reality at the same magnitude as taken by Aristotle. With the knowledge known at the time of Ancient Greece, Aristotle did almost the impossible into his Critical Thinking and Inquiry. Philosophy today is lacking this same amount of Inquiry, and there are so many more things to contemplate than at the time of Ancient Greece. Technology all around us should inspire a greater wave of new ideas, and thought experiments, and that needs to be fulfilled so that the World Civilization can benefit the most from these newly discovered concepts about the Universe.

Computer Language: AND, OR, XOR

All applications in Computers, and the compilation of all the data be it Numbers, Letters, Pictures, Videos, or Sound are stored and processed with the Binary Code 0s and 1s from the work of Transistors as on and off switches. The Language of Computers is Binary and follow the process of Logic Gates that can be AND, OR, XOR:

p	q	$p \vee q$	p∧ q	p→ q	p o q
T	T	T	T	T	F
T	F	T	F	F	T
F	T	T	F	T	T
F	F	F	F	T	F

In the table above, T represents a 1 and F a zero. In these Logic Gates, two inputs are sent, and Transistors are used, which then sends an output based on the Inputs. All possible answers are shown in the table.

^ stands for AND, v for OR, and O for XOR. P and q are the inputs.

Logic Gates:

The Logic Gates work as pathways leading to the diverse functions of a Software and Electronic Processing of Computers. In the same way that Brain Neurons interact with one another inside a Brain, the Transistors in a Computer work together Electronically allowing the Thinking Process of the Machine. Below is an example of pathways:

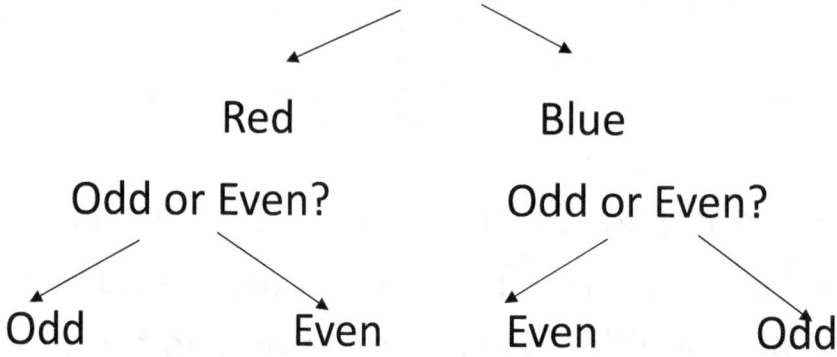

Welcome! Do you choose blue or red?

Red   Blue

Odd or Even?   Odd or Even?

Odd   Even   Even   Odd

Each question works as a Logic Gate that leads to a different Path. The Computer has Billions of Logic Gates to make a Software work.

Types of Transformers:

STEP DOWN TRANSFORMER:

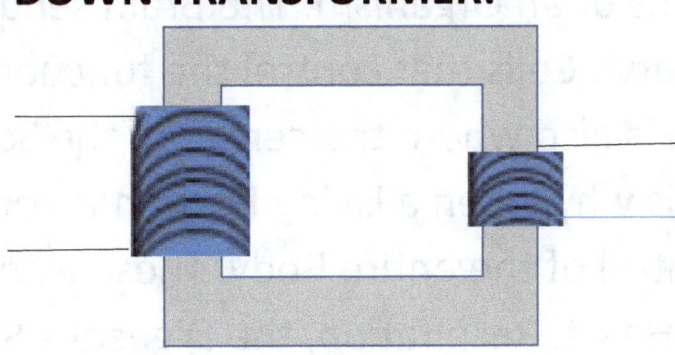

STEP UP TRANSFORMER:

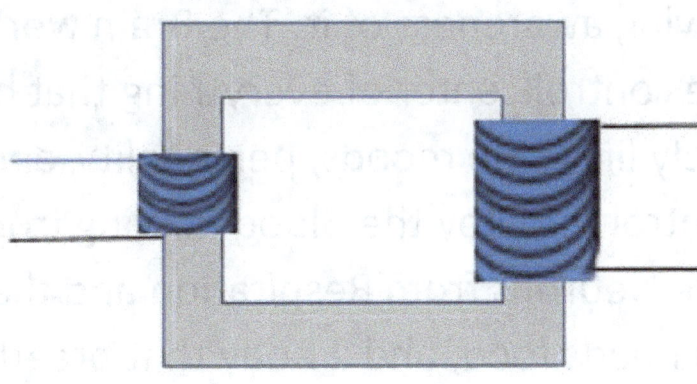

A Transformer is composed of a metal that has two sides with coils around it. A Step Up Transformer takes an Input Voltage in the coil on the left and Amplifies the Voltage in the Second Coil on the right. A Step Down Transformer does the opposite by decreasing the Voltage in the Second Coil. It Obeys the Following Equation:

$$\frac{Vs}{Vp} = \frac{Ip}{Is} = \frac{Ns}{Np}$$

V, I, and N are the Voltages, Current, and Number of turns in the coil. P and S stand for Primary and Secondary Coils. The Power IV on both coils are kept the same. The main goal of Transformers is to Amplify or decrease Voltage and control the flow of charges in a wire and they work on the basis of Electromagnetic Induction.

The Brain as the Computer of an Organism:

The Brain is one of the organs of an Organism. The Brain sends Electrical Signals through Nerve Cells that control the function of all the parts of the Body. The Brain is also where the center of the Soul resides in a Physical Body and that is why when a Living Being has control over the Brain, he or she has control of the entire Body. These Nerve Pulses regulate the beating of the Heart, Respiration, the Digestive System, any Physical Movement, Thinking Process, and many other things without the Living Being having awareness of it. The Brain works automatically and is like the control center of everything that happens to an Organism, and is closely linked to moods, personality, and behavior. This Center of Control is fed by the Blood flowing from the Heart, sending Energy to the Neurons from Respiration and the Digestive System. The Brain needs food, and a body that breathes Air so that it can acquire Energy for its Functions. A Body can exist without the Brain, but the Brain can not exist without the Body.

The Hearing:

Similar to a Microphone, the hearing of animals converts the vibration of air molecules into Electrical Signals that are sent to the Brain to be interpreted by the Observer. There is no hearing if there is no Brain, since that would be the same as a Microphone that is not connected to a device to record the Sound, or to emit these Sounds through a Speaker. In the same way if a person is not awake, that person can't hear the noises unless the noise is so loud that causes the person to awake from sleep. The Organisms are a Biological Machine that can be understood through the Laws of Physics and Chemistry at the Molecular and Quantum Level. In the same way that a person can be marveled by how the hearing works and the connection between a Person's Brain to the rest of the Body, when we open a Computer or gaze at a Car's Engine we can be equally marveled except that the Organisms came from nature and the Car or Computer are made by Humans. We are then truly capable of great marvels, and even possibly create our own Living Forms in Laboratory. Everything is within reach of our understanding. That is the belief that led us to the Computer and Artificial Intelligence.

Flow of Blood and Homeostasis:

There are Cycles everywhere in the Universe and there are Cycles inside all Living Beings. Photosynthesis in Plants follow a Cycle inside Chloroplasts and the Vegetal exchange with the surrounding Environment. All Living Beings are connected to each other as can be seen from the fact that without the Vegetal Life, Animals would lose a major source of Energy and would not be able to keep their Homeostasis. The Equilibrium of every Organism relies on its Respiration, which involves absorbing Energy from the surrounding be it Air, Water, Sunlight, or Food. Once Energy is acquired, many Chemical Processes occur that leads to the stability of the Living Organism that is Alive. These Cycles inside all forms of Life are also highly dependent of every Structure in the Universe, since the Sun and the Stars are a major source of Energy, and even Gravity is needed for Organisms to prosper in a Planet, Moon, or Asteroid.

Gears of the Universal Clock:

Our Galaxy moves in Space towards other Galaxies and away from other Galaxies. Our Sun moves around the Center of our Galaxy. Our Planet moves around the Sun leading to the Four Seasons of the Year. Our Moon moves around our Planet leading to the 12 Months of the Year. Our Planet rotates on its own axis leading to the 24 Hours of the Day. Cycles are repetitions that are found everywhere in the cosmos. The Electrons are found around the Nucleus of Atoms in Probability Clouds. The Universe is composed of gears that move other gears in the entire Universal Machine that is like an immense Clock. The Universe and the Earth undergoes Ages in these cycles leading to all History in Space Time. Time can't exist without Space, and Space can't exist without Time. In the same manner, the Electric and Magnetic Fields are intertwined in Light that propagates in Space. All things are entangled in the great Cosmic Web. Nothing is in isolation, and there are spins, rotations, and motion all around converting Energy to Matter, and Matter into Energy.

Technology and creation of truly living Organisms in Lab:

Given the amount of complexity seen inside Computers, Car Engines, Refrigerators, and Cell Phones, there is evidence that Science can truly be able to comprehend all things in the Universe. The number of people that understand how a Cell Phone works is very small, and this shows that the majority of the world population is ignorant of the current Scientific Understanding of the Natural World. A research on our modern Technology and the Thinking Process found in Computers revels that all Laws of Physics, and Understanding of the Cosmos is within human reach including time travel. Nothing will prevent scientists of the future from designing their own created life forms in laboratory, and from maneuvering Genes in the DNA, with the creation of more intelligent humans and other life forms that can live for 1000s of years without ever getting sick. The same revolution that led to our current Technology has the potential to lead to the creation of Super Organic Matter which is much more efficient than Artificial Intelligence found in Robots. If we can create a truly smart being made of flesh and blood like we, then we can do anything and even reach immortality.

Greek Philosophy: Aristotle

Evidently if Aristotle or Plato were alive today, they would engage in a deep Philosophical Inquiry, which would be based on all the advances in Science, and our current understanding of Space Time Reality, and our Modern Day Technology. There are several Mathematical Terms that fulfill a purpose as to improve human comprehension of the concrete patterns in the Cosmos. Could it be that the Universe is a Fractal Geometry and is infinite in extent? Can Fourier Series with its infinite sums lead to Finite material form in the Cosmos? Is matter really made of waves that when added together gives it form and its Physical Characteristics? Is the Cosmos a Mathematical Wave Function or Packet comprised of a large number of Wave Functions? What is Infinity according to Aristotle, and what exactly holds reality in place? In the next few pages I discuss more on Aristotelean Philosophy.

Infinity and Forms:

Aristotle argued that Infinity is not a substance and is not made of matter. The idea relies on the fact that a substance can be divided and is made of parts, but infinity can never be divided and is not comprised of parts. No two infinities can be bigger or smaller than each other. When stating that something is infinite, that makes two infinities just as infinite when compared to each other. Infinity is not an object, it is not made of matter but is possibly empty with a nature that is rather obscure. Infinity is what relies beyond the observable universe. He then explained that nothing contains infinity or else it would be infinite itself. Infinity alone is infinite. If infinity were made of parts, each of these parts would also be infinite which violated the concept that infinity is infinite already and no two infinities are more infinite than another. He also said that Infinity is not only not made of matter but it is also stationary and fixed, being immovable.

If infinity were to move it would move across something bigger than itself which is not allowed, so infinity is held fixed and at rest. Infinity is not an object or made of objects or else it would be possible to count infinity which being infinite is not countable. Infinity can not be compared to large numbers since infinity is boundless and anything boundless can not be compared to anything that is bounded. Things are complete only if they have an end or boundary. Something is considered the whole if it is bounded and includes everything. Completeness relates to size but Infinity has no form since it is boundless. What is unknowable can not contain and be determined which means that Infinity can't be grasped through reasoning. What is one can not be divided or else it would not be one. Time and movement are infinite since objects in the universe always move and will move. A number is never infinite since it is a finite amount. All things have a cause.

There can not be an infinite amount of causes or else it would not be possible to arrive at single thing. If A came from Z, which came from Y, which came from W, and an infinite number of causes, it would not be possible in that infinity to arrive at A. If there are an infinite number of causes that leads to A it would not be possible to ever arrive at the object A. There must be a point of start such as a Big Bang that created the universe. There is no before the Big Bang and there is only the after the Big Bang. Nothing that is infinite can be drawn with a finite line.

Aristotle said that space is where matter is. If there is an object in one location, space then obtains the shape of the object at the object's location. This statement from Aristotle brings to mind my ideas about dark matter, the ether, and the structure of the universe. Here is what I said:

According to my theory, matter is compressed space, and emptiness is stretched space. When space gets compressed matter and energy is formed, and when space is stretched there is nothing. When matter travels through space it uses new space as part of its composition. Space then goes right through the matter entering it which is why it is so hard to detect space or the Ether Wind simply because there is no Ether Wind. Space then is something, and truly nothingness do not exist.

Aristotle stated that there is nothing outside the All or the Whole. The Earth is in water, the water in air, the air in the ether, and the ether in heaven. The heaven is not anything else and all things are in heaven. Not everything that is in place is a movable body. Air is potentially water, and water potentially air but they are still water and air unless their potential state becomes actually such as in phase change, evaporation and condensation.

Aristotle said that space then must be something. Nothing in space is itself since its parts are something else. A liquid inside a jar is in a jar. The jar contains the liquid but is not the liquid. The jar and the liquid are formed in the space that gives them shape and form. Neither the liquid nor the jar are in themselves but are in something else. Space is like a river. The water in the river flows but the river itself is stationary. Space is fixed and does not move, but obtains the form of the object that is in it.

Aristotle did not believe on the Platonic Idea of forms. Plato's Idea of the forms stated that everything in the world is only a copy of something beyond this world in the realm of forms. Aristotle however observed that things in the world are too complex to just be a copy, and seem to be in themselves simply themselves. If we describe an object as a mere copy of its true essence in the world of forms we disregard the fact there must be several forms in the world of forms in order to give shape to all the forms in the world we live. If in the world of forms there is a man, and if all men were a copy of this man in the world of forms, then all men on earth would be equal to each other. Since each man in the world is different from each other then that implies that there must be just as many men in the world of forms in order to generate just as many men in our world to include all the diversity that we see. This statement is a challenge to Plato's Idea of the world of forms since now we got two worlds at the same size, and why not just have one?

Aristotle believe that all things are not one, but are just one to itself. All men are not one but each man is himself. He believed that things in themselves are too complex and the world of forms fails to explain the great complexity of all things. In order to explain the complexities, the world of forms would require so many information and why not spend time understanding what we see rather than refer to the forms beyond space and time every time?

Aristotle believed in what he could see and that was it. The natural world is already very complex and too much to be constantly mentioning the ideas in the world of forms which eventually fails since things are not copies but are unique in their existence. To understand what drives objects in our world to move becomes complex when we constantly refer to a higher form, while natural science should be based on the facts seen in our world and we must attempt to understand it through observation and the reasoning process shall be enough and should be the only way allowed.

How to reconcile Plato and Aristotle allowing both philosophies to live harmoniously with each other? Here is how:

Aristotle used Rhetoric to make Plato appear wrong, but Rhetoric can be used to support any cause whether right or wrong. It always narrows down to being able to speak and using evidence to support a cause even if a crazy cause. Anything is possible with Rhetoric, even to disprove that the world is round. There will always be a way to convince people that it is flat.

Socrates was put on trial accused of perverting minds. The knowledge of Rhetoric may lead someone to power, convincing people to follow him or her, leading to revolutions, and these were the fears of the Athenian State regarding Socrates. When a person grows too much in influence and power and leads to fear among those who in the past were the reference and the rulers of the society but who now see their power and influence in danger.

Thanks to Socrates, however, Greek thought reached several generations of thinkers for centuries, and a stimulus of thought that inspires people from all over the world to question things and have an appetite for wisdom. We know of Socrates through Plato's Writings, and Plato was also the teacher of Aristotle.

There is one absolute truth and not many that explains the universe:

Plato in one of his books wrote a conversation in which the absolute truth is one and truly universal, and truths are not relative to a person's opinion. Truth is not an opinion, but it is true whether a person agrees with it or not.

Truth:

The Sophists believed that what is right for a person is correct for that person, and what is right for another person is right for that other person. The problem with that statement as Socrates stated among Plato's Writings is that this idea could lead to anarchy and chaos since there is not a one single truth that everybody will follow.

A society can only work and become prosperous while functioning harmoniously between its constituents if everyone is under a single rule, principle, and where everyone follows a single truth. The Acceleration due to Gravity on the Surface of the Earth is 9.81 m/s^2 whether a person agrees with it or not that is an Absolute Truth. The fact is that there is one and only one Absolute Truth that rules all of the constituents of the Universe among this sea of Quantum Fluctuations. Truth is not biased but is rather the truth everywhere in the cosmos. The Truth is not an option or a preference but subdues everyone. If you jump off a cliff you will fall even if you want to fly, you will fall unless you have wings or a rocket of some sort. The Laws of Physics can not be broken unless on extremely rare cases, in which we call it a miracle. The Laws of Physics are the way they are and no thought process or opinion can change it. In other words: The Sophists were wrong, and their ideology is dangerous and Socrates knew that too. Sophists ideology has the potential to bring chaos the fall of society since every person follows its own rule, which simply can't work.

In the next pages, there is the story of Electrons travelling in a Circuit becoming stored information in the Binary Format of Computer Language:

The Electronic Trip in the Grid comprised of Capacitors and Transistors.

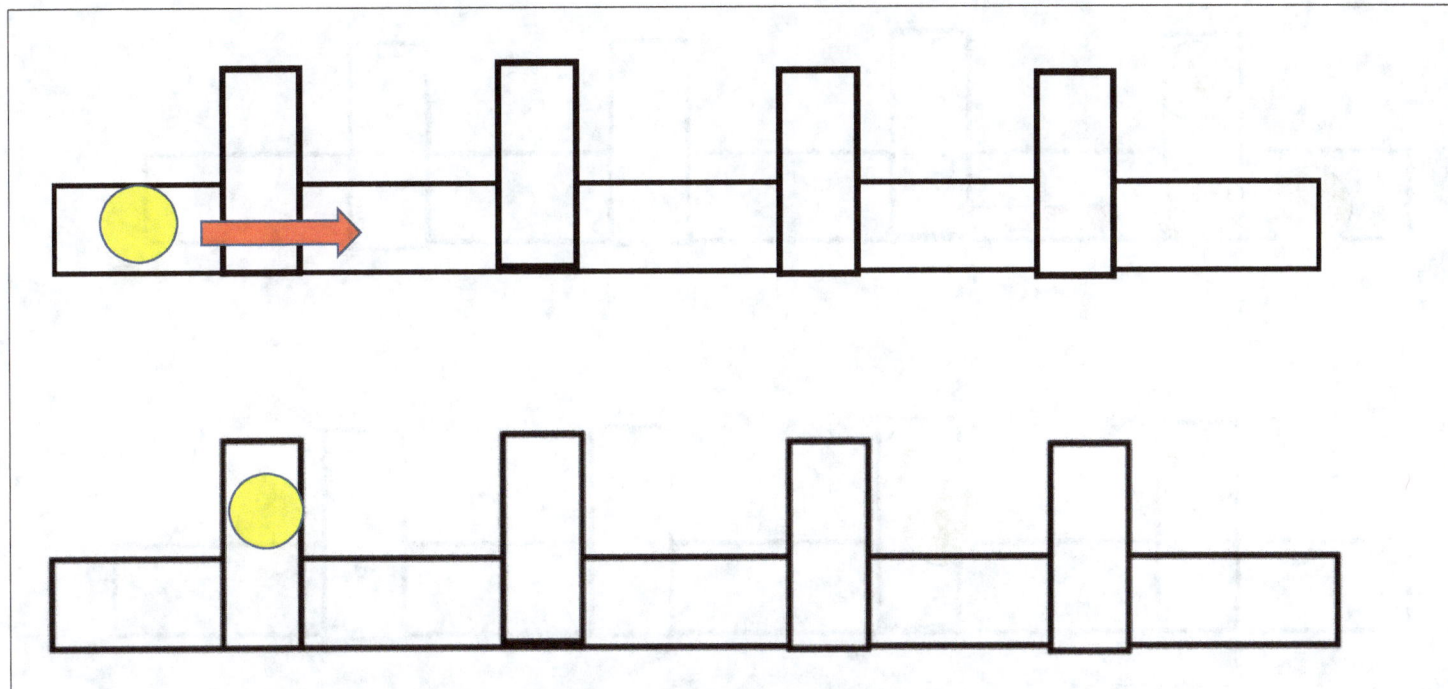

An Electron comes in and enters the first slot.

Each Slot is a Capacitor.

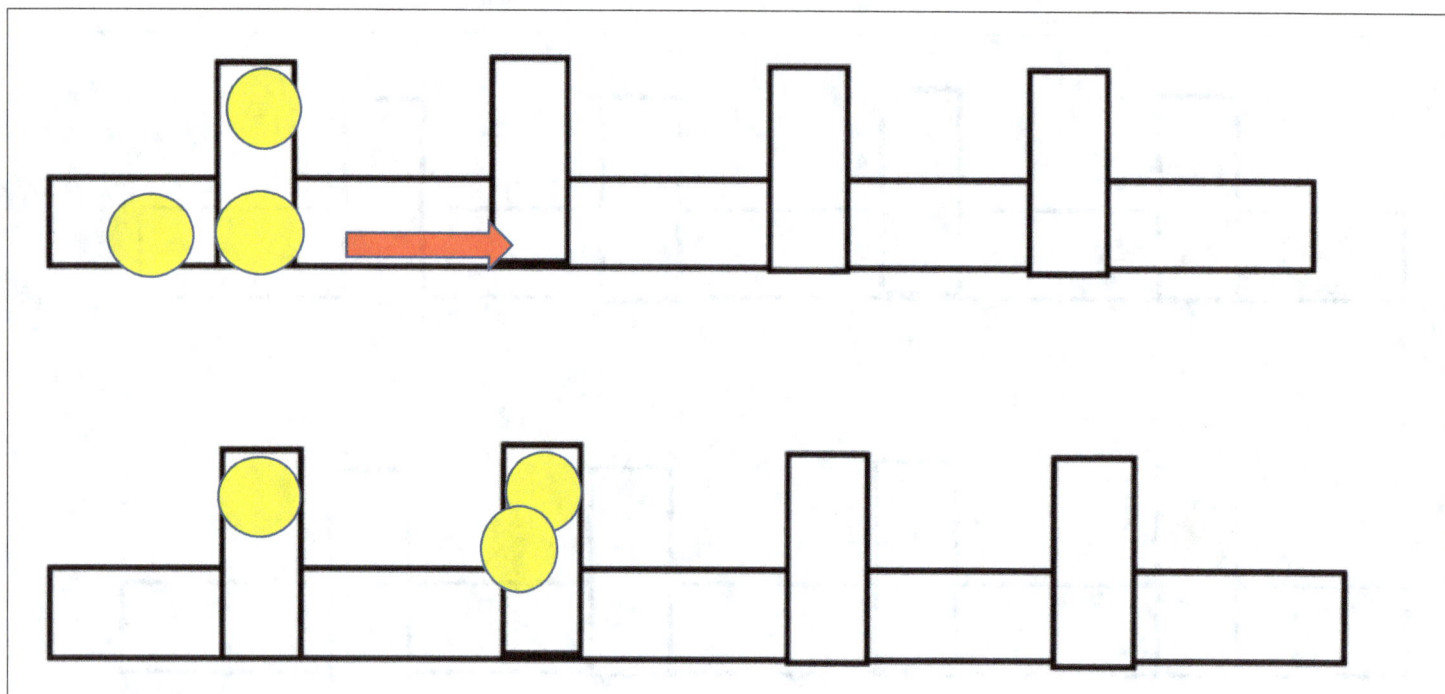

Two Electrons come in are repelled by the first
slot and enter the second slot.

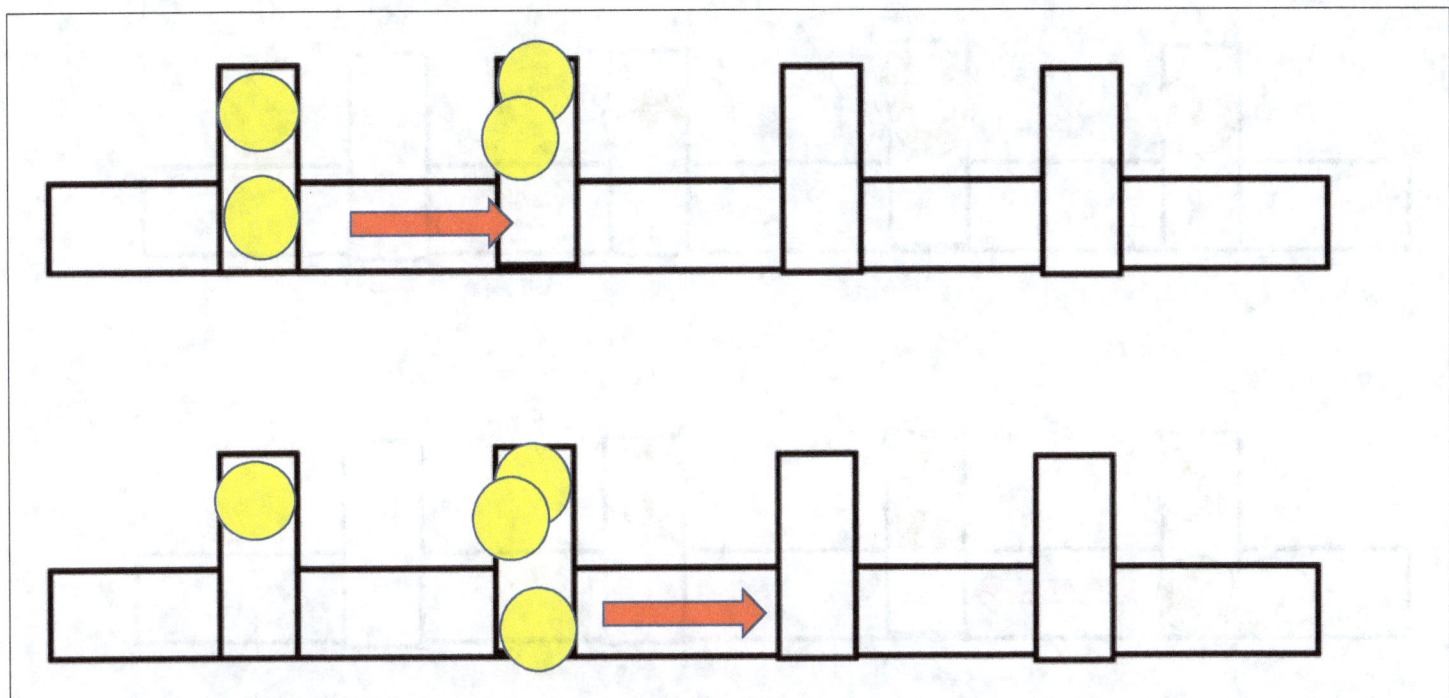

One Electron comes in, is repelled by the first and second slot and in the next page this Electron enters the third slot.

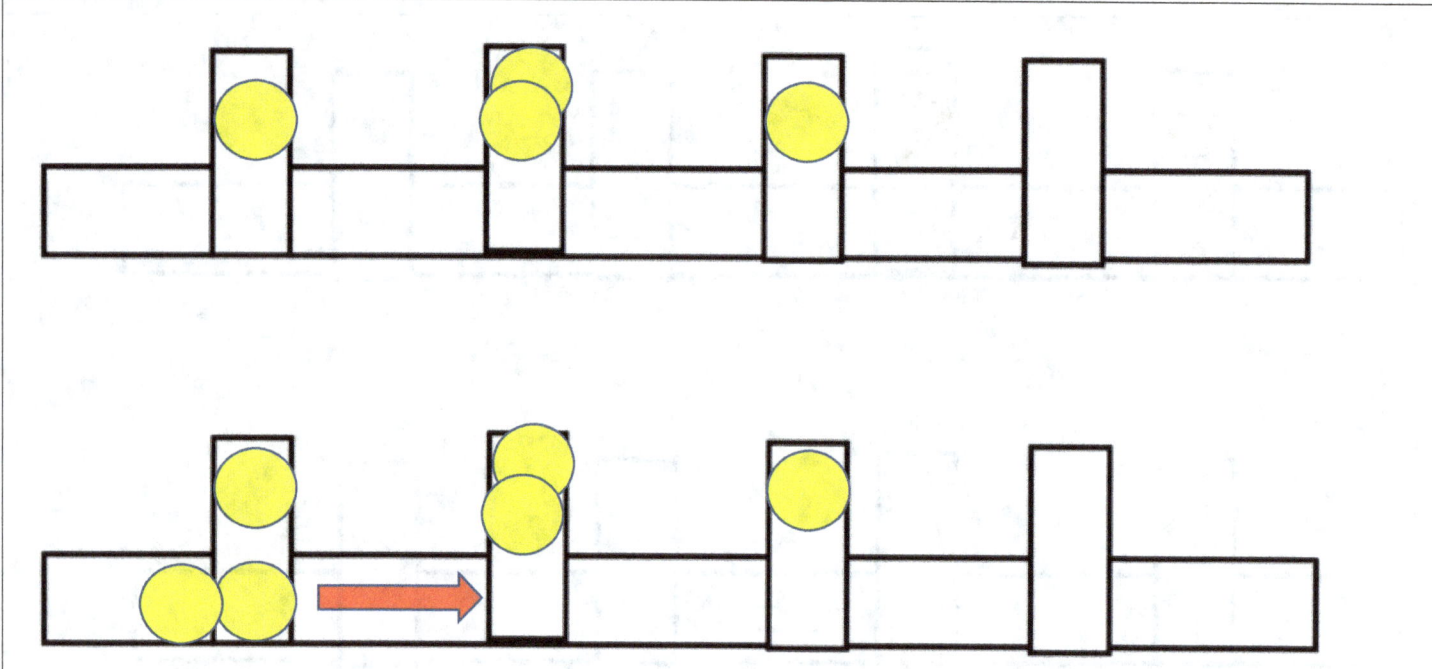

Two Electrons come in and are repelled by the first, and
in the next page they are also repelled by the second
and third slot.

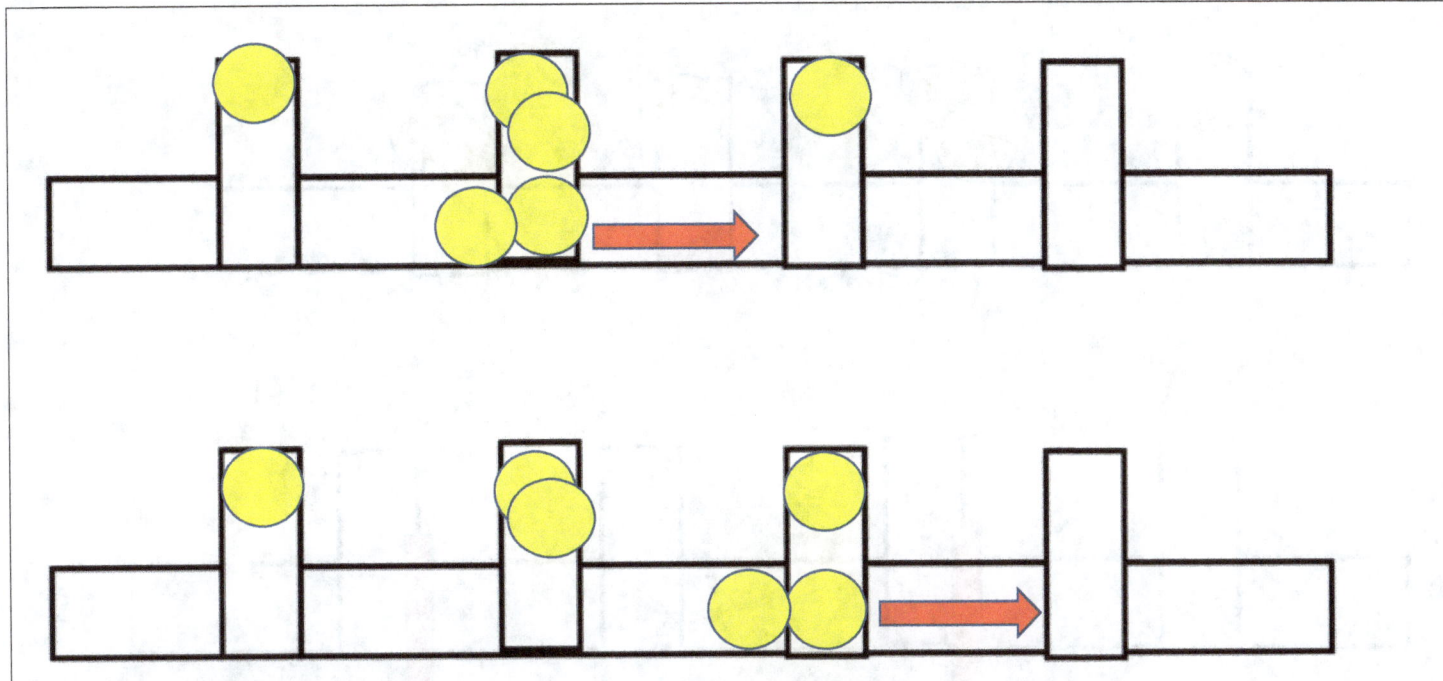

These two Electrons then enter the fourth slot that had no Electrons.

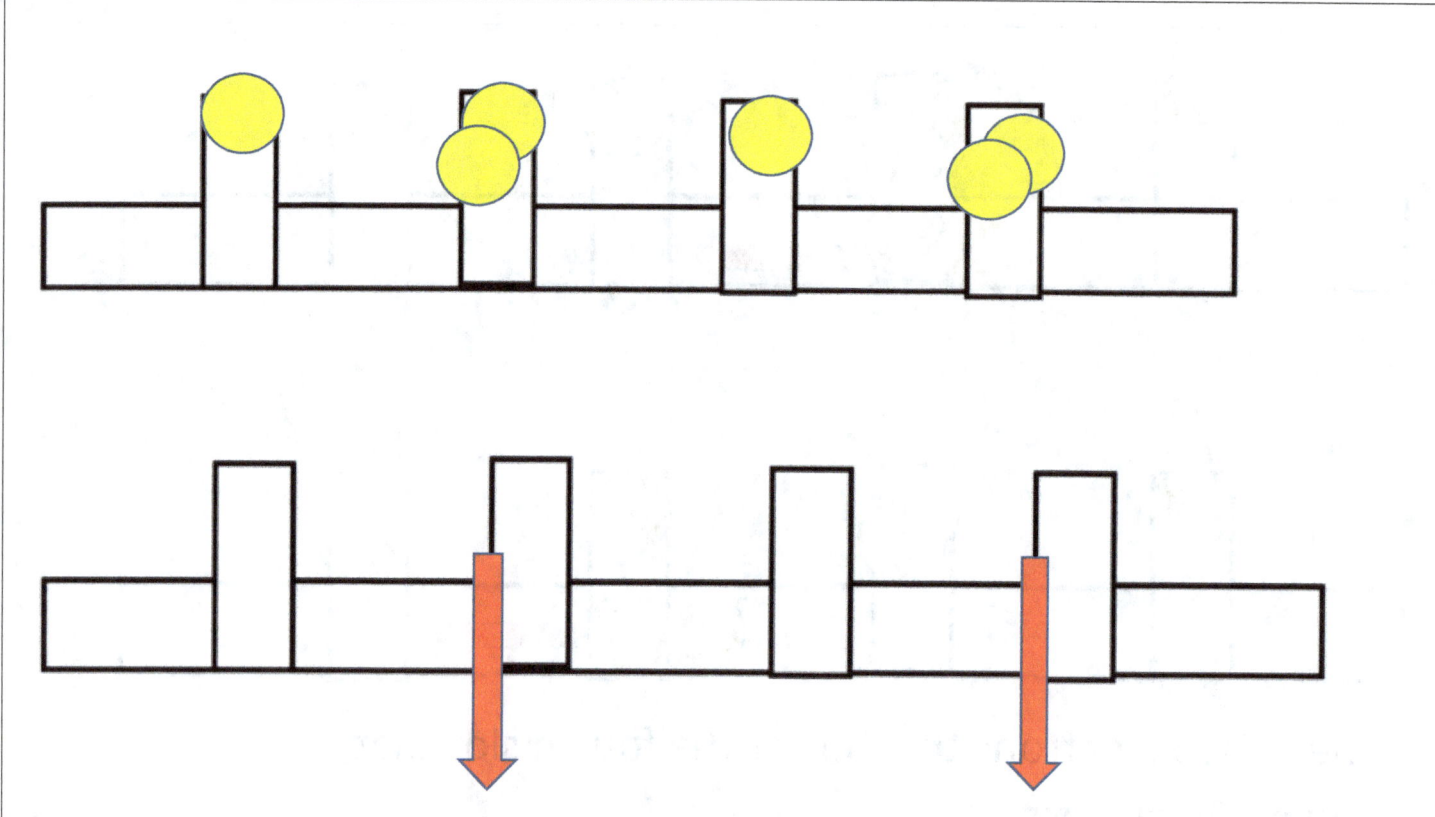

The slots with one Electron represents a 0 and the slots with two Electrons represent a 1. Their charges are released by Transistors with the Binary signature 0 1 0 1.

Convergence and Fourier Series:

In Fourier Series a periodic function is taken and an infinite sum of sinusoidal waves is performed which converges to that periodic function. Fourier Series can be used to demonstrate sum of waves which compose matter and energy in the universe. Take for example a table which is made of a superposition of wave functions that when added together makes the form of the table. Space itself is a sea of wave functions which are the great amount of particles and energy constantly flowing. We consider space a Hilbert space.

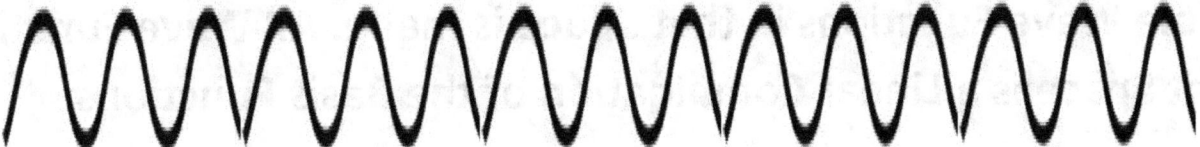

Hilbert Space Problems:

Imagine a Hilbert Space composed of Numbers 1, 2, and 3.

All Waves are composed of a combination of these three numbers in this particular Space. Here are the possible Wave Functions in this example:

1 alone, 2 alone, 3 alone, 1 + 2 = 3, 1+ 3 = 4, 2+3 = 5, and 1 + 2 + 3 = 6.

These are the possible outcomes of Wave Functions in that Space:

1,   2,   3,   4,   5,   and   6   all six of these Functions use the numbers 1,2, and 3 as their Basis, and they are all the possible outcomes in this Hilbert Space.

The Basis is a set of Wave Functions in Space from which all Possible Wave Functions in that Space is made. All Wave Functions in that Space is a Linear Combination of the Basis Functions.

Hilbert Space Probabilities:

Suppose there is a Particle with the following Quantum States:

1 1 1 2 2 3 3 3 3 4

There are 10 numbers.

The probability of a Particle in being in the Quantum State 1 is:

3/10 since there are 3 number 1s.

The probability of a Particle in being in the Quantum State 2 is:

2/10 since there are 2 number 2s.

The probability of a Particle in being in the Quantum State 3 is:

4/10 since there are 4 number 3s.

The probability of a Particle in being in the Quantum State 4 is:

1/10 since there is 1 number 4.

3 + 2 + 4 + 1 = 10

All the possible Quantum States have their probabilities add to 100%. Some Quantum States are more or less likely than others but in which Quantum State the Particle is in is a mystery. Without being observed the Particle exists in a Wave of Probabilities and when observed the Particle appears to be in only one Quantum State. This is the Basis for the Uncertainty Principle still not very well understood in Physics.

We live in a Mathematical Universe that Mathematicians desire to have it measured in Discrete Amounts but in which Mathematicians themselves fall prey to the Probabilistic Aspect of many things in the Natural World. Not all things can be 100% determined, and Quantum Mechanics proves that with Scientists using Statistics to understand the existence and State of Particles. Also with Human Development, wisdom comes first in a person's life and only later comes maturity. Maybe Probabilities come first in Quantum Mechanics but only later will we fully grasp its meaning.

Maturity and Experience:

Aristotle stated that abstract knowledge of Mathematics and Geometry comes first in a person's life when still young, but right judgement and decisions in life are more complicated and require experience. Virtue then takes time to be reached, and reason alone can't do it, but must be accompanied with experience only gained after a long a life. Quantum Mechanics is still a very young Science and is not yet a Complete Theory and neither is known by more than 0.1% of the World Population.

The Planck Length:

The Planck Length is the smallest length in Space possible. The Planck Length are like building blocks which when added together generates the apparent smooth Space and Time. The Universe can not be smaller than the Planck Length. The Planck Length is a single Pixel of all the Pixels in the Universe. The Universe can not be infinitely small since that defies the quantum concept that Space is comprised of an enormous number of small pieces and nothing can be smaller than those pieces. The Singularity at the moment of the Big Bang could not have been smaller than the Planck Length either. The size of the Universe where the Theory of Relativity and Quantum Mechanics are unable to reconcile is the limit from which the Universe can not be smaller than that particular size. That is why it is not possible for Quantum Mechanics and the Theory of Relativity to be in agreement with each other at such small scales simply because the universe can never be that small. The Planck Length is the limit size of the Universe at the Singularity before the Big Bang.

The Golden Ratio:

A   B

$$\frac{A+B}{A} = \frac{A}{B} = \phi = 1.618....$$

Equation to help solve for possible values of A and B

$$A = B(1.62)$$

$$B = A/(1.62)$$

The Longer Part divided by the Smaller Part is equal to Whole Part divided by the Longer Part.

Golden Ratio

5/3 = 1.67

8/5 = 1.6

13/8 = 1.63

Pascal's Triangle

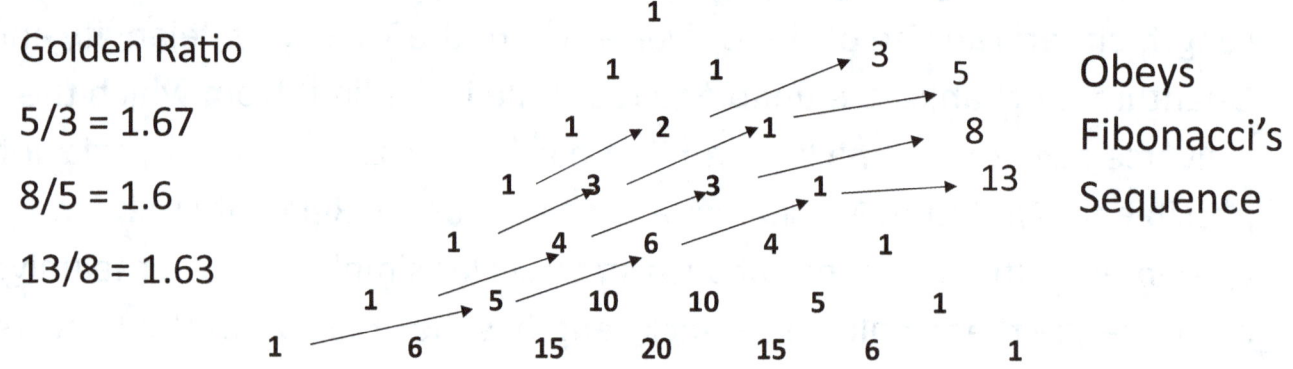

Obeys Fibonacci's Sequence

Golden Rectangle:

1   1   2   3   5   8   13   Sequence seen in Pascal's Triangle

This Ratio or Proportion is seen in many parts of the Natural World being considered a Fine Proportion for Beauty.

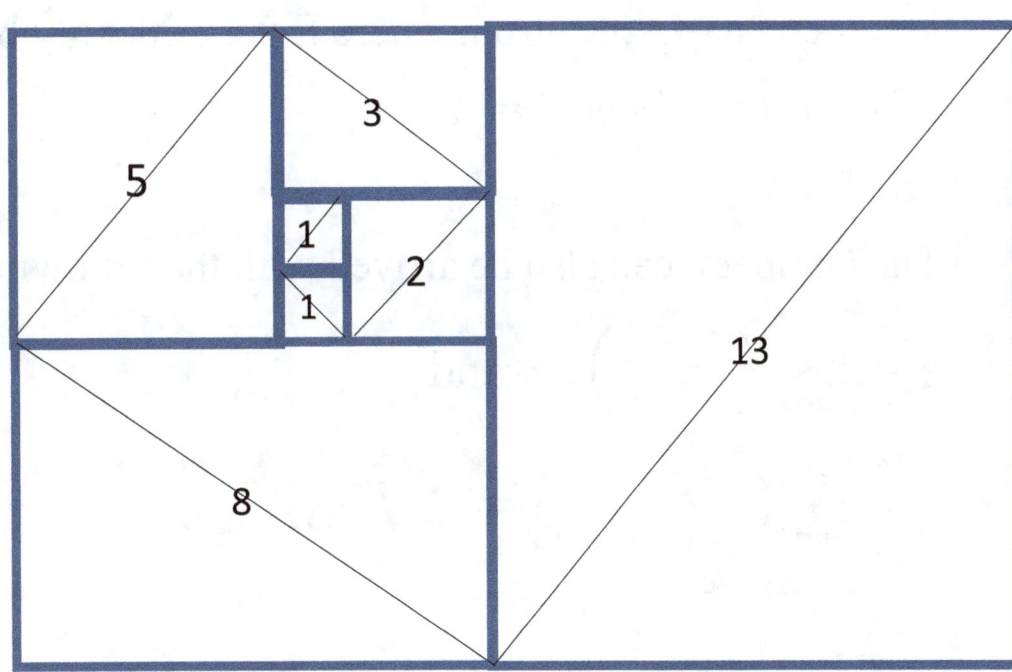

The Value of e:

The Derivative of e^x is itself. The Integral of e^x is also itself. The area under a curve from negative Infinity to positive Infinity of e^{-x^2} is equal to $\sqrt{\pi}$. Using Taylor Series to expand Sine, Cosine, and e^{ix}

We arrive at the following equation:

$e^{ix} = cosx + isinx$ from where Euler's Identity becomes:

$e^{i\pi} = \text{Cos}(\pi) + i\text{Sin}(\pi) = -1 + 0$

So $e^{i\pi} + 1 = 0$

The number e can also be arrived with the following equations:

$$e = \lim_{n \to \infty} \left(1 + \frac{1}{n}\right)^n \quad \text{and}$$

$$e = \sum_0^\infty \frac{1}{n!} \qquad e = 2.718281\ldots\ldots$$

Imaginary Number i:

$$\int_{-\infty}^{\infty} (e^{ix})^{ix}\, dx = \sqrt{\pi}$$

i stands for an Imaginary Number which is used as a Mathematical Trick or Tool in order to solve equations which is needed in Science especially Physics. Whether Imaginary Numbers exist or not in Nature is a mystery, but their presence is found in many equations of Quantum Mechanics, Probability, and Statistics. Its usage is evidence that something like Imaginary Numbers does in fact exist in the structure of the Natural World. i is also related to the number e and π in equations such as the one shown above. All three numbers are truly transcendental and appear everywhere in Mathematics.

$$i = \sqrt{-1}$$

Pi: π = 3.14159……

Pi is how many times the Circumference is bigger than the Diameter of a Circle. This number contains infinity within the circle since it is not a rational number and has decimals that extend throughout infinity with no clear pattern. Truly mysterious.

π can be approximated by regular polygons of ever greater number of sides towards infinity ever closer to a perfect circle but never a perfect circle.

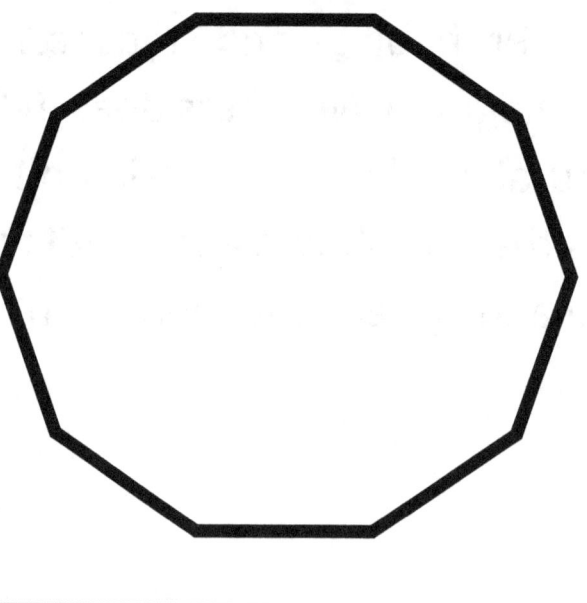

The Infinity of π is possibly evidence that a perfect circle does not exist.

The Integral: $\dfrac{1}{\sqrt{\pi}} \int_{-\infty}^{\infty} (e^{ix})^{ix} \, dx = 1$

Shows that the Area under that Gaussian Curve is exactly equal to 1. The Graph of the Function is shown below:

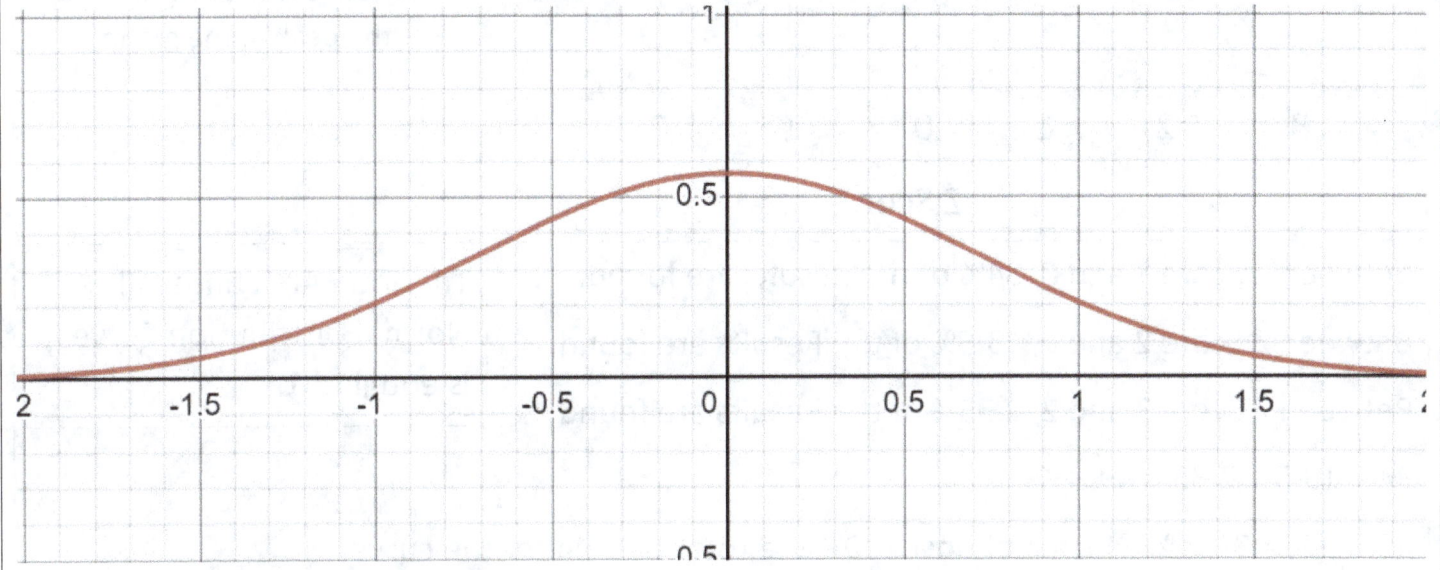

1 is also a very unique number representing Unity, Completeness, and Wholesome.

Normal Distribution:

Z Score

The Normal Distribution Graph is a Gaussian Curve with the x axis being the Z score which indicates how far away is a specific result from the mean which is at the peak of the graph.

Between Z Score 1 and 0, 34% of the Results are found.

Between Z Score 2 and 1, 13.5% of the Results are found.

Between Z Score 3 and 2, 2.35% of the Results are found.

34 + 13.5 + 2.35 = 49.85%

49.85(2) = 99.7 % which is almost 100% the total area under the curve.

The Standard Deviation is a measure of how spread is the data relative to the mean. Each Z Score is one Standard Deviation from the Mean. A Z Score of 2 is 2 Deviations from the Mean.

The Total Area under all Normal Distribution Curves is equal to 1.

Fine Structure Constant: 1/137

Major examples of where it can be found:

1...The Ratio of the Speed of the Electron in the first orbit of Hydrogen to the Speed of Light.

2...Relates to the Probability that an Electron will emit or absorb a Photon.

3...The difference in Energy between two Electrons of different Spin that are resolved in the Spectrum.

There are many other cases where the Fine Structure appear in Physics which makes this number truly especial in all Natural Phenomena. Physicists are unable to understand why the Fine Structure exists but calculations show that if the Fine Structure Constant and many other Constants were slightly different, the cosmos would not have been able to form and life would be impossible.

a = 1/137

The Fine Structure Constant sets the strength of the Electromagnetic Force since it is the probability that and Electron absorbs or emits Photons, and that is the way in which Electromagnetic Force propagates in space. Electric Fields are generated by the emission of Photons.

Since this is a constant found within Electromagnetism, it is of no surprise that it appears in many calculations of Quantum Mechanics:

$$a = \frac{e^2}{4\pi\varepsilon_0 \hbar c} \qquad \Longleftarrow \qquad a = \frac{\left(\dfrac{e^2}{4\pi\varepsilon_0 d}\right)}{\left(\dfrac{hc}{\lambda}\right)}$$

4...The Ratio of the Electric Potential Energy between two Electrons to the Energy of a Photon of Wavelength equal to the distance between these two Electrons with $\lambda = 2\pi d$.

5...The Ratio of the Potential Energy U_e of the Electron in the Ground State of the Hydrogen Atom to its Rest Mass $m_e c^2$.

$$a^2 = \frac{U_e}{m_e c^2}$$ The Ratio equals the Fine Structure Squared in this example

6...The Maximum Number of Protons in an Atom that will make a single Electron move in an orbit at the Speed of Light. Z = 137.

$$\frac{mv^2}{r} = \frac{Ze^2}{4\pi\varepsilon_0 r^2}$$ Z is the Number of Protons

In the same way that h appears everywhere in Quantum Mechanics, it is not surprising that a common ratio such as 1/137 seen in atomic coupling will likewise. The only difference is that 1/137 is a quantity with no units.

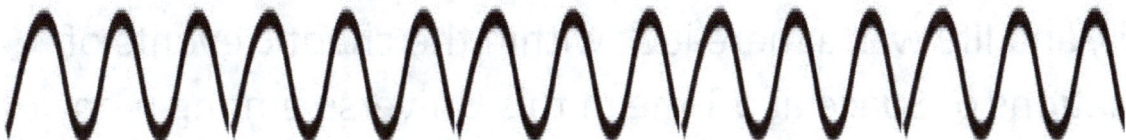

Anthropic Principle:

The Anthropic Principle is the idea that if the Universe had different Laws and Constants that were not favorable for life, no human being would exist to state the imperfections of these laws into not forming life. If the Universe were a little different, there would be no person alive to state that the cosmos are not perfect. Humans only exist because the Physical Laws and Constants were favorable for life, and had these Laws been different no person would be around to contemplate reality. Possibly, there are multiple Universes with all sorts of Laws that are not capable to sustain Life Forms. We just happen to live in one of the many Universes that has the right Laws and Constants. The Universe may not exist with the purpose to form humans. The Universe probably exists blindly, and life was a mere luck within the chaotic events of Fluctuations of Space and Time in this Universe among many others that failed to bring the birth of Life. Our Universe was just lucky and won the lottery. Universes with life are probably rare.

What are Constants in Nature?

Constants exist within the simulation that molds the Space Time Reality that we live in. The same Physical Rules that exist on Earth, exist on Mars, Venus, in the far away Stars, and in the farthest Galaxy. Constants exist in Equations that lead to the results by matching the proportions of the relations between individual values of variables that lead to a result that is in agreement with experiments or observations. It is possible to known the relationship between variables and only then discover the value of a constant that can translate the proportions into a numerical result that matches experimental values. There are many Constants among multiple Equations of Physics and they are the signature found in all Equations connecting all the parts of the Universe that truly obey these Constants and Laws like a machine or computer following directions. The Constants is the Language that is responsible for organizing and for the functioning of all Natural Phenomena and the one reason we exist to contemplate about this reality.

DIFFERENTIAL EQUATIONS:

Here begins a short investigation on the Science of Rate of Change with respect to variables.

$$\frac{dy}{dx} \qquad \int_{x1}^{x2} F(x)\,dx$$

Example 1:

$$\frac{dy}{dx} = \frac{7lnx}{xy}$$

$$ydy = \frac{(7lnx)dx}{x}$$

$$\frac{y^2}{2} = \frac{7(\ln(x))^2}{2} + c$$

$$y = \sqrt{7(lnx)^2 + c}$$

To solve this problem the y portion of the equation is separated from the x portion and the integrals are taken individually. The solution is then set equal to y.

Example 2:

$$xy' - 4y = 4x^4 lnx$$

$$y' - \frac{4y}{x} = 4x^3 lnx$$

$$e^{\int \frac{-4}{x} dx} = e^{-4lnx} = x^{-4} \text{ the integration factor}$$

$$y'x^{-4} - 4x^{-5}y = \frac{4lnx}{x}$$

$$yx^{-4} = \int \frac{4lnx}{x} dx$$

$$\frac{y}{x^4} = \frac{4(lnx)^2}{2} + c$$

$$y = x^4(2(lnx)^2 + c)$$

Here, an Integrating Factor is used to cancel x^4 in $4x^4 lnx$

and then both sides are integrated and the results are solved for y.

Example 3:

$$(y^2 + 5xy + 4x^2)dx - x^2dy = 0$$
$$(y^2 + 5xy + 4x^2)dx = x^2dy$$
$$\frac{(y^2+5xy+4x^2)}{x^2} = \frac{dy}{dx}$$

$$y = xv \qquad y' = x\frac{dv}{dx} + v$$

$$\frac{(x^2v^2+5x^2v+4x^2)}{x^2} = x\frac{dv}{dx} + v$$

$$v^2 + 5v + 4 = x\frac{dv}{dx} + v$$

$$v^2 + 4v + 4 = x\frac{dv}{dx}$$

Here, the y and x portions of the problem are separated and a substitution such as the

$$y = xv \qquad y' = x\frac{dv}{dx} + v$$

is used to then lead to an easier calculation of the integration of both x and y, and the solution is at the end solved for y.

$$(v + 2)^2 = x\frac{dv}{dx}$$

$$\frac{dv}{(v + 2)^2} = \frac{dx}{x}$$

$$\int \frac{dv}{(v + 2)^2} = \int \frac{dx}{x}$$

Before solving for y, the results reverse the substitution method so that the answer can be equal to y rather than x or v.

$$\frac{-1}{(v+2)} = \ln x + c$$

$$-(v + 2) = \frac{1}{\ln x + c}$$

$$-v - 2 = \frac{1}{\ln x + c}$$

$$-\frac{y}{x} = \frac{1}{\ln x + c} + 2$$

$$y = -x(\frac{1}{\ln x + c} + 2)$$

Example 4:

$(y - x^2)dx + 3xdy = 0$

$My = \dfrac{d}{dy}(y - x^2) = 1$

$Nx = \dfrac{d}{dx}3x = 3$

When $My \neq Nx$

$\dfrac{My-Nx}{N} = \dfrac{1-3}{3x} = \dfrac{-2}{3x}$

$e^{\frac{-2}{3}\int\left(\frac{1}{x}\right)dx} = e^{-\frac{2}{3}lnx} = x^{-\frac{2}{3}}$ The integration factor

$(yx^{-\frac{2}{3}} - x^{\frac{4}{3}})dx + 3x^{\frac{1}{3}}dy = 0$

Here, the equation is evaluated to see if My = Nx, which in this case is not. The Integrating Factor from $\dfrac{My-Nx}{N}$ is used and applied on the equation and now My = Nx with both equal to $x^{-\frac{2}{3}}$.

$$\int 3x^{\frac{1}{3}}dy + f(x) = c$$

$$3x^{\frac{1}{3}}y + f(x) = c$$

$$Mx = x^{\frac{-2}{3}}y + f'(x)$$

$$f'(x) = -x^{\frac{4}{3}}$$

$$f(x) = -\frac{3}{7}x^{\frac{7}{3}}$$

$$3x^{\frac{1}{3}}y - \frac{3}{7}x^{\frac{7}{3}} = c$$

$$3x^{\frac{1}{3}}y = c + \frac{3}{7}x^{\frac{7}{3}}$$

The Integral of the y portion of the equation is performed in terms of y. A derivative with respect to x is applied to that equation and compared with Mx. In order for the integral with respect to y to be equal to Mx, $f'(x)$ has to equal $-x^{\frac{4}{3}}$, which means that

$f(x) = -\frac{3}{7}x^{\frac{7}{3}}$. The value is then placed in the equation which is solved for y.

$$.y = \frac{c + \frac{3}{7}x^{\frac{7}{3}}}{3x^{\frac{1}{3}}}$$

Example 5:

$$ydx + (-3x - y^5)dy = 0$$

$$My = 1$$

$$Nx = -3$$

$$My \neq Nx$$

$$\frac{My - Nx}{M} = \frac{1+3}{y} = \frac{4}{y}$$ The integration factor

When we use M instead of N in the denominator to calculate the integration factor we do:

Here, My is not equal to Nx, so an Integrating Factor is used is order so that they can be equal. Since the M is at the bottom of the equation for the Integrating Factor, a negative sign is placed in front of the Integral.

$e^{-\int \frac{4}{y}}$ with a negative sign in front of the $\int \frac{4}{y}$ instead.

$e^{-\int \frac{4}{y}} = e^{-4 \ln y} = y^{-4}$

Plugging it in we get:

$y^{-3}dx + (-3xy^{-4} - y)dy = 0$

My = Nx now

My = $-3y^{-4}$

Nx = $-3y^{-4}$

$\int y^{-3}dx + g(y) = c$

$y^{-3}x + g(y) = c$

Ny = $-3y^{-4}x + g'(y) = c$

After the Integrating Factor My = Nx. The Integration of the x portion is made with respect to x and the derivative is set equal to Ny. In order for them to be equal, g'(y) = -y, which means that $g(y) = -\frac{y^2}{2}$.

g'(y) = -y

$$g(y) = -\frac{y^2}{2}$$

$$-3y^{-4}x - \frac{y^2}{2} = c$$

Example 6:

$$\frac{3}{2}\frac{dy}{dx} + \frac{3e^x y}{1 + e^x} = 3\sqrt{y}e^{-x}$$

Divide the whole equation by $\sqrt{y}$

$$\frac{3}{2}y'y^{-\frac{1}{2}} + \frac{3e^x y^{\frac{1}{2}}}{1+e^x} = 3e^{-x}$$

With u = $y^{\frac{1}{2}}$

The results are set to equal y as the solution of the equation.

On Example 6, the entire equation is divided by $\sqrt{7}$ and a u substitution is used together with u' to put the equation in a quadratic format.

and $u' = \frac{1}{2}y^{-\frac{1}{2}}y'$

We get:

$$3u' + \frac{3e^x u}{1+e^x} = 3e^{-x}$$

Divided by 3:

$$u' + \frac{e^x u}{1+e^x} = e^{-x}$$

and:

In this format, the equation can be solved with an Integrating Factor. The factor is then multiplied throughout the equation and Integration is done on both sides of the equal sign.

$$e^{\int \frac{e^x}{1+e^x}dx} = (1+e^x) \text{ The integration factor}$$

$$u'(1+e^x) + e^x u = e^{-x}(1+e^x)$$

$$u(1+e^x) = \text{ the integral of } (e^{-x}+1)dx$$

$$u(1+e^x) = -e^{-x} + x + c$$

$$u = \frac{-e^{-x}+x+c}{1+e^x}$$

$$\sqrt{y} = \frac{-e^{-x}+x+c}{1+e^x}$$

$$y = \left(\frac{-e^{-x}+x+c}{1+e^x}\right)^2$$

An inverse substitution is then made and the equation is solved for y.

Example 7:

$$y''' + 4y'' - 5y = 0$$
$$(r^3 + 4r^2 - 5)=0$$

The solutions for r above is:

$$r = 1$$

$$r \approx -1.4$$

In example 7, the quadratic equation of derivatives is solved with roots and translated into an answer with the number e on it.

$r \approx -3.62$ The answer is then: $y = c1e^x + c2e^{-1.4x} + c3e^{-3.62x} + c$

Example 8:

In example 8, the roots of the derivatives is solved and when the roots repeat such as in this case, a variable x is placed in front of the terms in the solution.

$(y" + 4y' + 4y) = 0$

$(r^2 + 4r + 4) = 0$

$(r + 2)^2 = 0$

Solutions are:

.$r = -2$ twice

So $y = c1e^{2x} + c2xe^{2x}$

When the r solution repeats we insert an x at the second repeated term.

Example 9:

The roots here are solved and because of the imaginary number sin and cos functions are incorporated in the answer.

$$y^{(4)} + 16y'' + 63\, y = 0$$
$$(r^4 + 16r^2 + 63\,) = 0$$
$$(r^2 + 9)(r^2 + 7) = 0$$

Solutions are:

$r = 3i$

$r = \sqrt{7}i$

So the answer is:

$$y = c1\cos(3x) + c2\sin(3x) + c3\sin(\sqrt{7}x) + c4\cos(\sqrt{7}x)$$

if the solutions were in the form of one complex and one real this is what we would get instead as shown below:

$$2 + 3i = c1e^{2x}\cos(3x) + c2e^{2x}\sin(3x)$$

Example 10:

Rules for Annihilator:

1) $cx^k e^{ax} = (d - a)^{k+1}$
2) $cx^k e^{ax}\cos(bx) = (D^2 - 2abD + a^2 + b^2)^{k+1}$
3) $x^k = A + Bx + cx^k$

These are the rules to solve Differential Equations following known mathematical patterns.

One example:

$$xsinx + e^{-2x}$$

b = 1 , a= 0, and k =1 for xsinx

a= -2, and k = 0 for e^{-2x}

So,

$$(D^2 + 1)^2(D + 2)$$

In example 11, the same is done and then a value for A is solve and placed inside the solution

Example 11:

In differential equations there may be homogenous and particular solutions.

$y'' - 8y' + 16\,y = 4e^{-2x}$

$(r - 4)^2 = 0$ homogeneous solution

Solution r = 4 twice

$(r - 4)^2(D + 2)$ whole solution

$y = c1e^{4x} + c2xe^{4x} + (Ae^{-2x}$ particular solution)

Solving for A:

$P = Ae^{-2x}$

$P' = -2\,Ae^{-2x}$

$P'' = 4\,Ae^{-2x}$

$4Ae^{-2x} - 8(-2\,Ae^{-2x}) + 16(Ae^{-2x}) = 4e^{-2x}$

4A + 16A + 16A = 4

36A = 4

$$A = 1/9$$

So the whole solution is homogeneous plus particular

$$y = c1e^{4x} + c2xe^{4x} + (1/9)e^{-2x}$$

Differential Equations is the Mathematics of the rate of change and finding solutions for equations in one or multiple variables. Mathematics is the study of numbers and patterns that can be represented with figures, geometry, and thoroughly understood with Trigonometry. All things are numbers, ad Mathematics is the language of all Science since all Atoms, Molecules, and Chemical and Physical Reactions involve ratios, rate of change, and proportions within known Laws and Constants.

Zyn Tablets of Inquiry

diogo de souza

This book is a journey through abstract pictures and a few words since pictures are worth a 1000 words. It is a book of reflection about the arts, science, and philosophy. Its intent is to lead the reader into thinking, and to inspire the artist into drawing his or her concept of the universe whether right or wrong it will not matter. Imagination is bigger than knowledge and much can be discovered when imagination is allowed to flourish freely.

Tablet

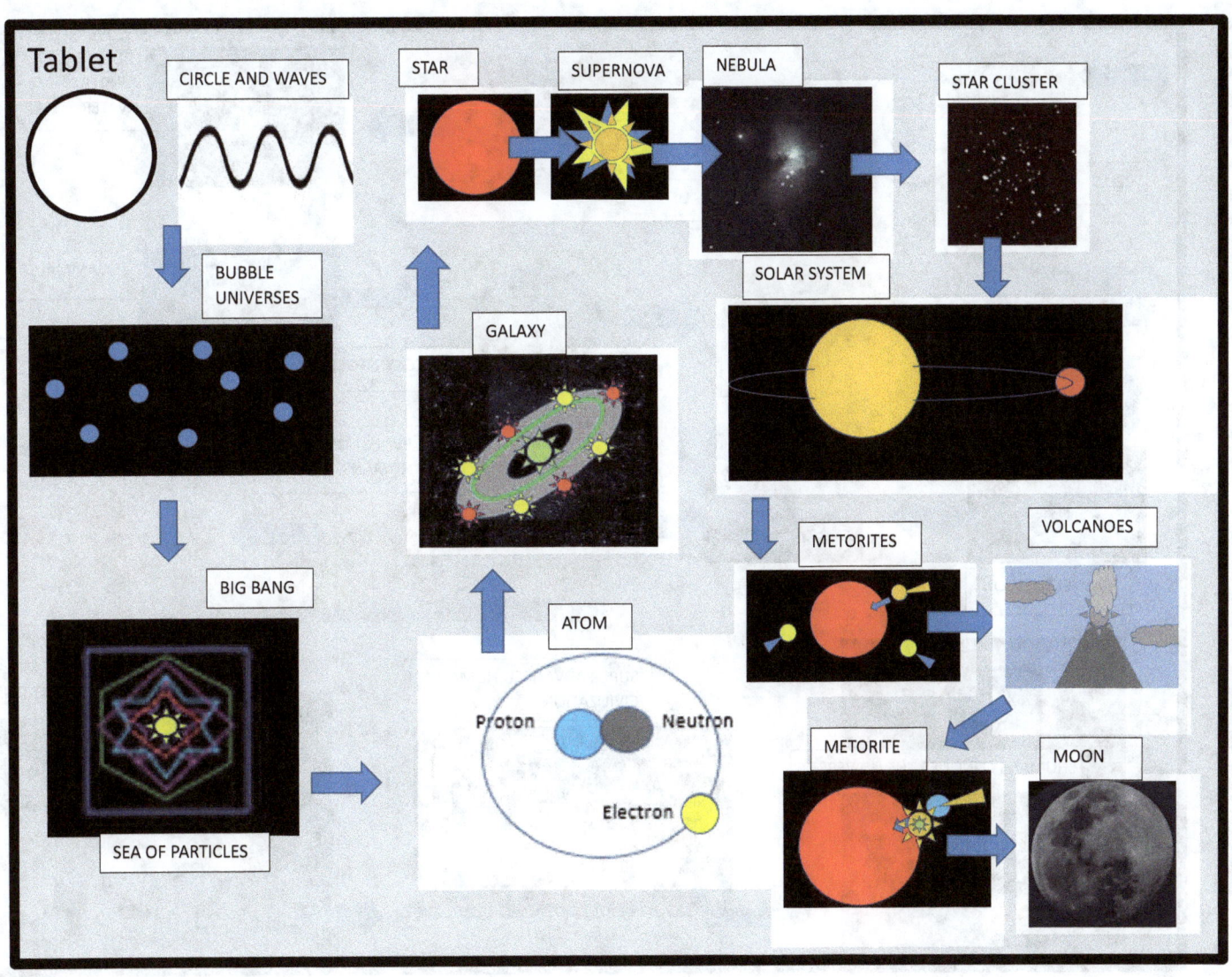

CIRCLE AND WAVES

STAR SUPERNOVA NEBULA STAR CLUSTER

BUBBLE UNIVERSES

GALAXY SOLAR SYSTEM

BIG BANG

ATOM METORITES VOLCANOES

Proton Neutron

Electron

SEA OF PARTICLES

METORITE MOON

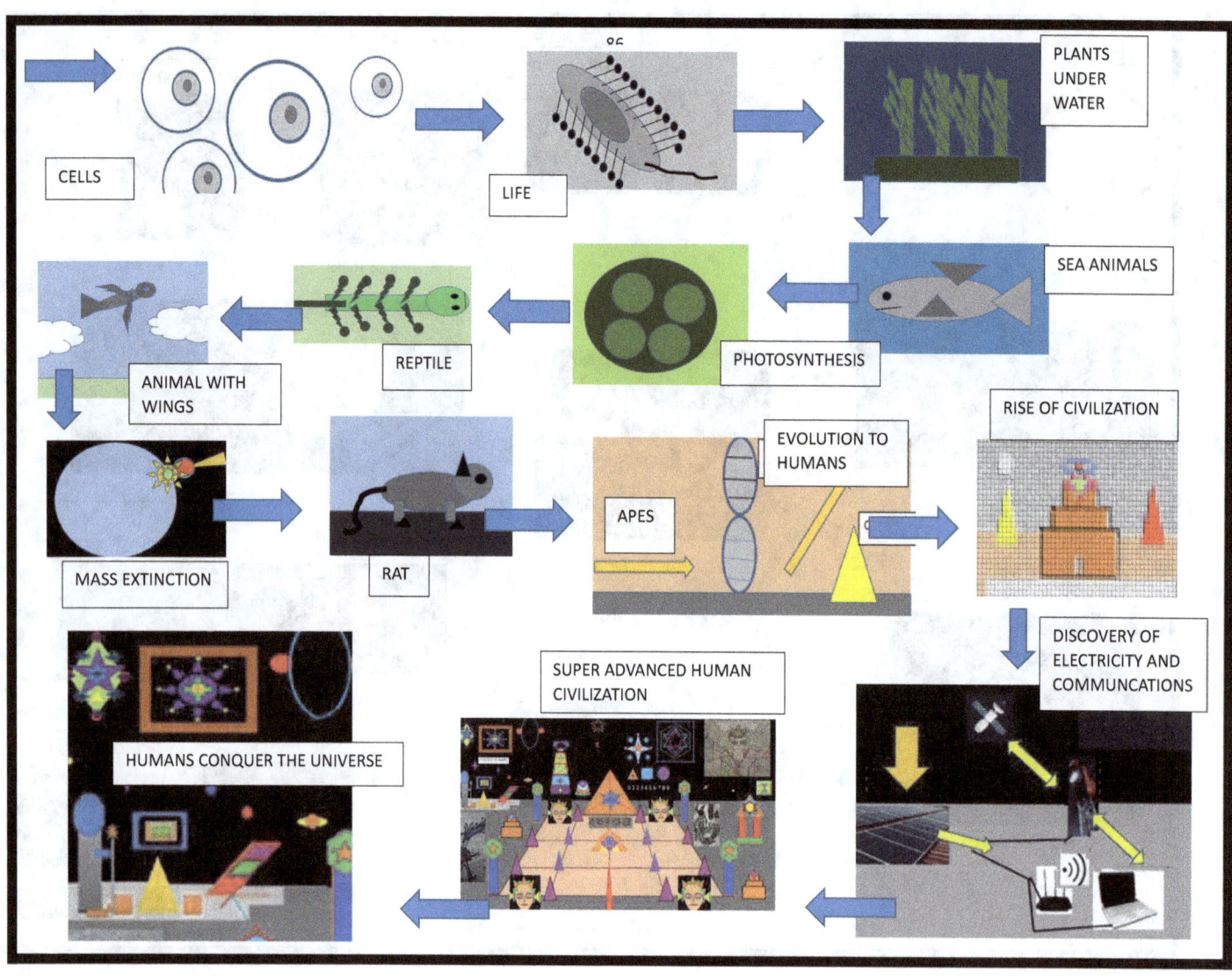

The first two pages show two Tablets with pictures and words with arrows. The arrows describe the Evolution of Time. Starting among the Quantum Fluctuations that gave rise to the Birth of the Universe, later the formation of Atoms from Particles bonding together as the Universe expanded. Then comes the first Stars beginning to shine, the formation of Galaxies, the Death of Stars that went Supernova, and then the Nebulas formed from that explosion, and new Stars emerging among the clouds according to the Conservation of Energy. Matter is created from pre-existing Energy. The Universe was created from pre-existing Quantum Fluctuations. The Solar System is then put together in those clouds that condensed, with Earth as a ball of Plasma being bombarded by Meteorites, and Volcanoes that erupted forming the Atmosphere. Life then emerges with the first Cells, and the first Simple Organisms in the Seas. Aquatic Plants, and Aquatic Animals, and Photosynthesis are generated. Then comes the Reptiles later the Dinosaurs and Animals with Wings. Mass Extinctions occur every once in a while, and one of them annihilated the Dinosaurs, but Rat like creatures survived, to later evolve into many animals, including the Apes, that later evolved into Hominid Creatures that became Human after a long Evolutionary Process. After many Thousands of Years, Human Civilization is constructed with Farming, Domestication of Animals, and Great Cities and Mass Production of Materials, as well as Economy, and Politics. Electricity is discovered and Humans advance beyond the Earth into Space, and then meeting their goal as an Intelligent Race of Beings when building a Perfect World highly advanced in Science and Social Order. How did all of that happen? This simple explanation is described in the next page. The answer for all things in the Universe relies on one word called:

TIME. Given enough time anything can happen.

IT TOOK 13.8 BILLION YEARS FOR A PLANET TO BE BORN AND TO GIVE RISE TO LIFE AND INTELLIGENT LIFE. THE UNIVERSE CREATES MATTER FROM ENERGY, BUT GIVEN ENOUGH TIME, INTELLIGENT LIFE MAY COME TO EXISTENCE IN A WORLD LIKE OUR OWN. GIVEN ENOUGH TIME ALL SORT OF ACCIDENTS AND MULTIPLE AND DIVERSE THINGS WILL OCCUR IN HISTORY.

DUE TO MANY STARS AND WORLDS IN THE COSMOS, THERE IS A HIGH CHANCE OF LIFE BEING BORN IN A PLANET, AND GIVEN THOUSANDS OF MILLIONS OF YEARS, HUMAN LIKE BEINGS COME TO EXISTENCE.

LIFE EXISTS ON EARTH FOR BILLIONS OF YEARS AND ONLY NOW RECENTLY HUMANS WERE FORMED BY NATURE. THINGS IN THE COSMOS TAKES TIME, AND IT IS A GAME OF CHANCE. AFTER ROLLING THE DICE LONG ENOUGH, INTELLIGENT LIFE MAY ONE DAY SHOW UP IN THE DICE BEING ROLLED TOO MANY TIMES.

NATURE ROLLS DICE. SUPPOSE A SIMULATION OF A UNIVERSE IS A HAND THAT ROLLS SEVEN DICE. EACH ROLL IS A YEAR SINCE THE BIG BANG. HOW MANY TIMES SHOULD THIS HAND ROLL THE SEVEN DICE UNTIL ALL OF THEM RECORD THE NUMBER ONE. LET US SUPPOSE THAT WHEN ALL SEVEN DICE SHOW A ONE AT THE SAME TIME, THAT IS WHEN INTELLIGENT LIFE IS BORN. IT MAY TAKE SEVERAL ROLLS AND SEVERAL YEARS BUT EVENTUALLY THIS WILL HAPPEN SINCE THE UNIVERSE LASTS FOR BILLIONS OF YEARS. A BILLION ROLLS. THE UNIVERSE ROLLS DICE. LIFE IS A CHANCE AND DUE TO MANY STARS, PLANETS, AND GALAXIES, THEIR GREAT NUMBERS INCREASED THE LIKELIHOOD THAT IN ONE PLANET SUCH AS EARTH, LIFE, AND EVENTUALLY INTELLIGENT LIFE WAS FORMED.

$$\left(\frac{1}{6}\right)^7 = \frac{1}{279936}$$

IN OUR SIMULATION WITH SEVEN DICE, IT MAY TAKE 279,936 ROLLS TO FINALY GET ALL ONES IN ALL THE DICES AT ONCE. THAT MEANS THAT IT WOULD TAKE 279,936 YEARS FOR INTELLIGENT LIFE TO BE BORN. WE CAN MAKE EACH ROLL OF DICE THOUSANDS OF YEARS INSTEAD OF ONE YEAR. THE POINT IS CLEAR HOWEVER, THAT BEING ETERNITY VERY LONG, AT ONE TIME IN THE FUTURE THE UNIVERSE WINS THE LOTTERY AND HUMANS ARE BORN.

HUMANS EXIST FOR THOUSANDS OF YEARS AND ONLY RECENTLY HAD THE TECHNOLOGY TO LAND ON THE MOON. GIVEN ENOUGH TIME, WITH HUMANS ROLLING THEIR DICE FOR TENS OF THOUSANDS OF YEARS, EVENTUALLY A LOTTERY IS WON AND HUMANS LEARN TO READ, WRITE, COUNT, BUILD CITIES,

DISCOVERS ELECTRICTY, AND THEN LANDS ON THE MOON. THINGS IN THE UNIVERSE HAPPEN IN SLOW MOTION BUT EVENTUALLY SOME PARTICLES SAY: BINGO! THE SAME ARGUMENT EXISTS FOR THE BIRTH OF LIFE, AND DUE TO SO MANY WORLDS IN THE ENTIRE COSMOS, OTHER PLANETS MIGHT SAY BINGO AND LIFE IS BORN. THE CHANCES ARE GREAT FOR OTHER LIFE FORMS IN OTHER PLANETS SIMPLY BECAUSE THERE ARE MORE THAN A TRILLION WORLDS OUT THERE.

The universe is large, and all things that can happen within this sea of probabilities does in fact happen. The greater the Telescope more new discoveries are made in the Cosmos. In this Space there are many floating worlds in a place that is complete dark with many points of light which are Stars, where Infinity and Eternity seem obscure. What was there before the beginning, and what will there be in all of Eternity? We can call the Cosmos our Family, but in this Family there are Trillions of Trillion Worlds. Too big and filled with danger, and diversity. Planets and Stars vary in sizes, and all things are complex enough and difficult to grasp. In the next page there is the map of the Universe with our Planet Earth in the Center.

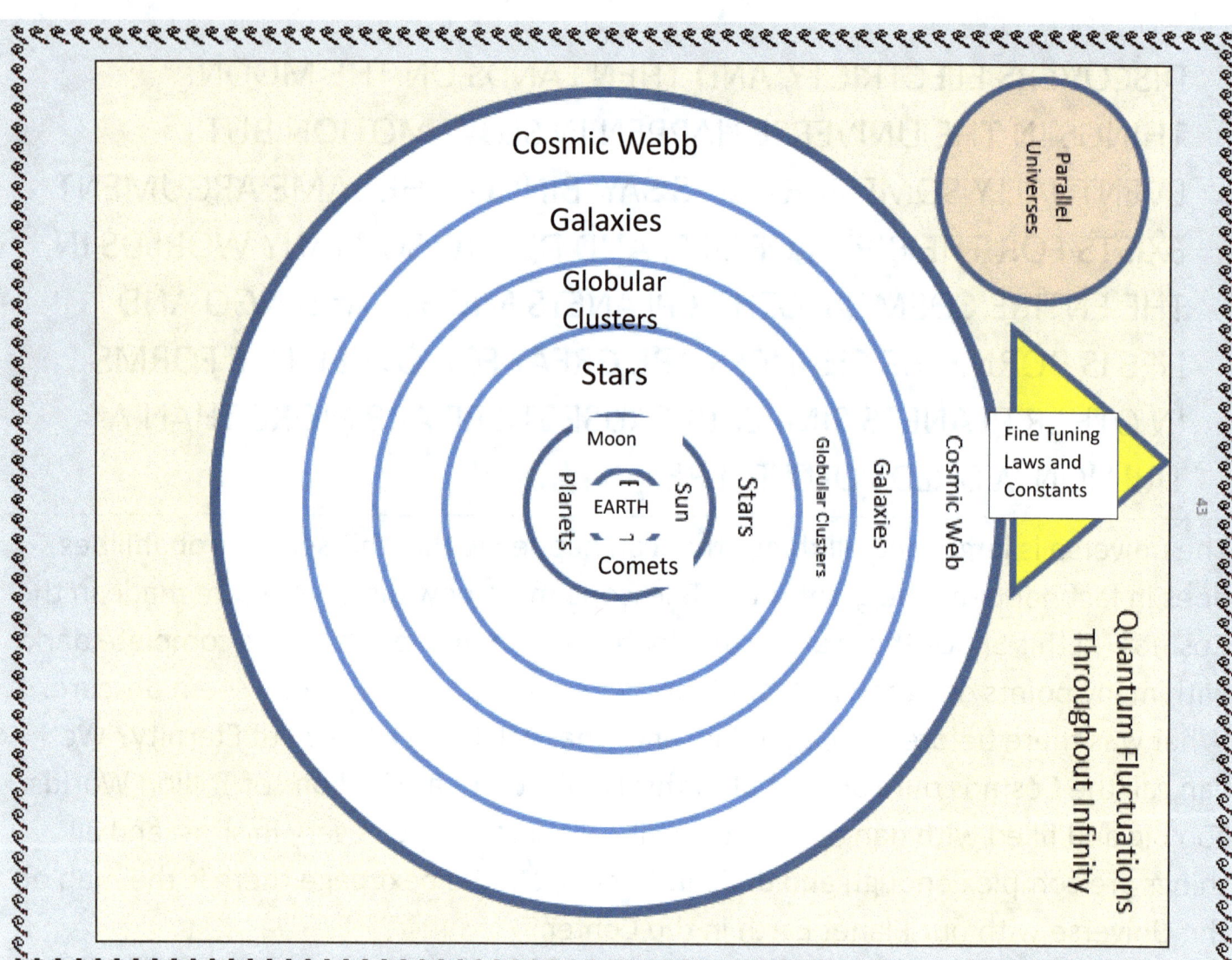

In the previous picture, if Earth is placed in the center of the known Universe: The Sun, Moon, and Planets appear close by orbiting the Earth, further away comes the Stars, further away the Globular Clusters, further away the Galaxies, and extremely far away is the Cosmic Web of Dark Matter connecting several Galaxies and Galaxy Clusters in the Structure of the visible Universe. Outside the Universe are the Parallel Universes, the Triangle represents the Fine Tuning in the Weak Anthropic Principle where we exist due to Laws and Constants that made that possible, and beyond all things are the Quantum Fluctuations that create several Universes and Particles throughout Infinity.

The Dice, the DNA, and Evolution:

Inside every Cell of every Living Being there is the DNA. The DNA is a very long and Complex Molecule comprised of a Row of Molecules bond to each other. It is truly a work of Art how the combination of these Molecules leads to the Traits, and the Body Functions of all Living Organisms. The fact is that these are the Mutations which lead to the Evolutionary Process and Natural Selection akin to the roll of Dice. The Blind Nature keeps rolling Dice all the time, until the Molecules say Bingo, and the Mutations lead slowly over Millions of Years, to new creatures, to new forms of Life.

In Space there is a Sea of Molecules Vibrating and movingly randomly. Over time in a period of Millions of Years, this Random Motion will eventually get something right which causes Evolution and the birth of New Creatures from Pre-Existing ones. Since the motion is at random, many creatures obtain less favorable traits and become extinct. The greatest proof that Nature is Blind and that it rolls Dice, is that most of the random Motion leads to mass extinctions. When rolling a Dice, it is more likely to get something wrong than something right. When something wrong happens which is more likely, the Living Being dies. Since Nature rolls Dice for so many years, that increases the likelihood of getting at least something right in the future. Something right will happen, however, without doubt, and one of the things that went right was when Apes evolved to Humans. We are now the Humans and the Evolutionary Process continue to happen. I wonder how many times nature will roll Dice getting something wrong, until it finally gets something right once again. Given enough time it will say BINGO!

Stargazing at the Multiplicity of Worlds:

The Stars in the Sky always fascinated humans who used the position of these Celestial Bodies at distinct times in a year to create a Calendar, by keeping track of many events, festivals, and religious ceremonies. The Ancient Civilizations were able to see that the Heavens appeared to rotate around the Earth making one full rotation a day, and one full rotation in a year. The Moon takes a Month to revolve around the Earth once, and this fact was used to establish the 12 Months in a Year. The 12 Constellations of the Zodiac also became associated with the 12 Months where Astronomers recorded the position of the Sun along the Ecliptic. The Sun makes one full rotation across the Zodiac through the Ecliptic in about 365 days which are separated in 12 Months. The Sun is seen to spend about a Month in each of the Constellations along the Ecliptic. The Ancient Astronomers also noticed that the Sky had a Line of Stars and dust running across the Celestial Sphere. They recorded that there were more Stars in this dust lane than elsewhere in the Sky. Today we know that this dust lane in the Sky is the Milky Way Galaxy that is shaped like a disc. The Solar System is part of the Milky Way and from the Earth we have a sideways view of this flat disc in the form of a dust lane or path in sky where the population of Stars are more concentrated. Unfortunately with the rise of the Industrial Revolution and the advent of Electricity, Light Pollution does not allow people living in urban areas ability to see this great view of the Stars and of our Galaxy. I consider this a major problem that is often overlooked but that it completely damages research of Space as seen from the Earth making it difficult to enjoy the sky with a Telescope.

THE HEAVENS THROUGH A SMALL TELESCOPE

Light pollution is the biggest enemy of Astronomy. In a truly dark sky away from city lights any small Telescope is capable of revealing wonders. Stargazing under heavy light pollution may require a DSLR Camera attached to the Telescope by taking advantage that these cameras are more sensitive to light than our eyes making it easier to stargaze that way. Some people may find boring not being able to see with their own eyes without the aid of a camera. In the next few pages are images from a DSLR Camera with one shot exposure of no more than 30 seconds at or near the DFW Metroplex Area in Texas.

STAR CLUSTERS

PERSEUS DOUBLE CLUSTER

M 35

GLOBULAR CLUSTER

HERCULES GREAT GLOBULAR CLUSTER

GALAXIES   ANDROMEDA

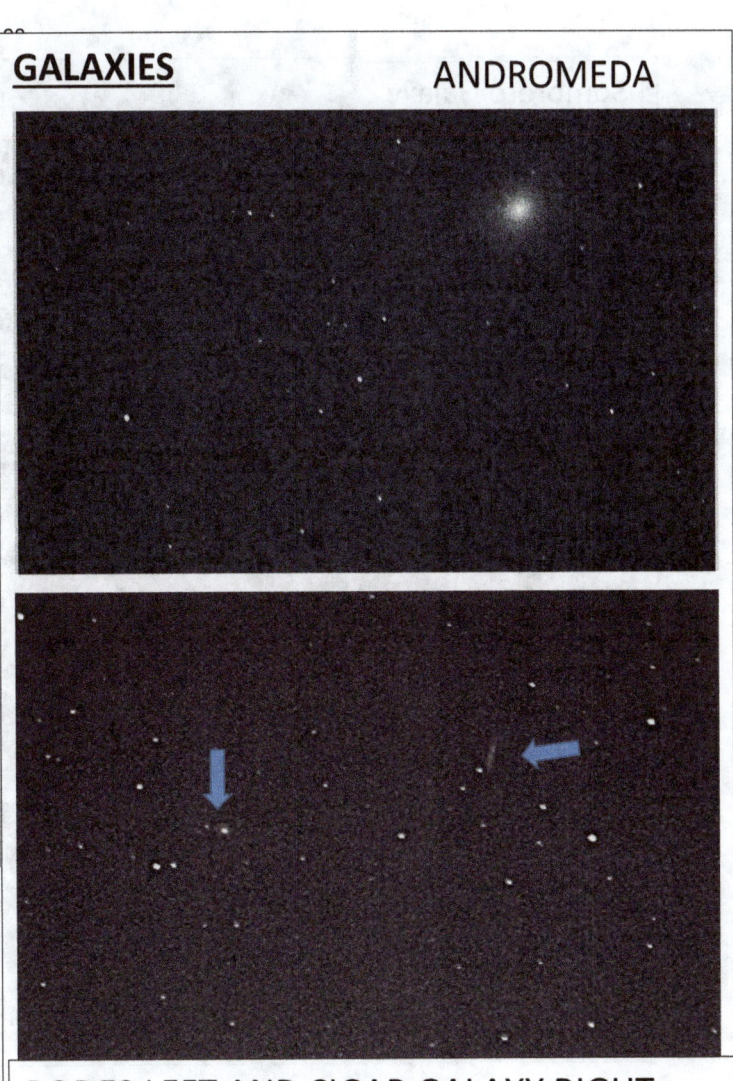

BODES LEFT AND CIGAR GALAXY RIGHT

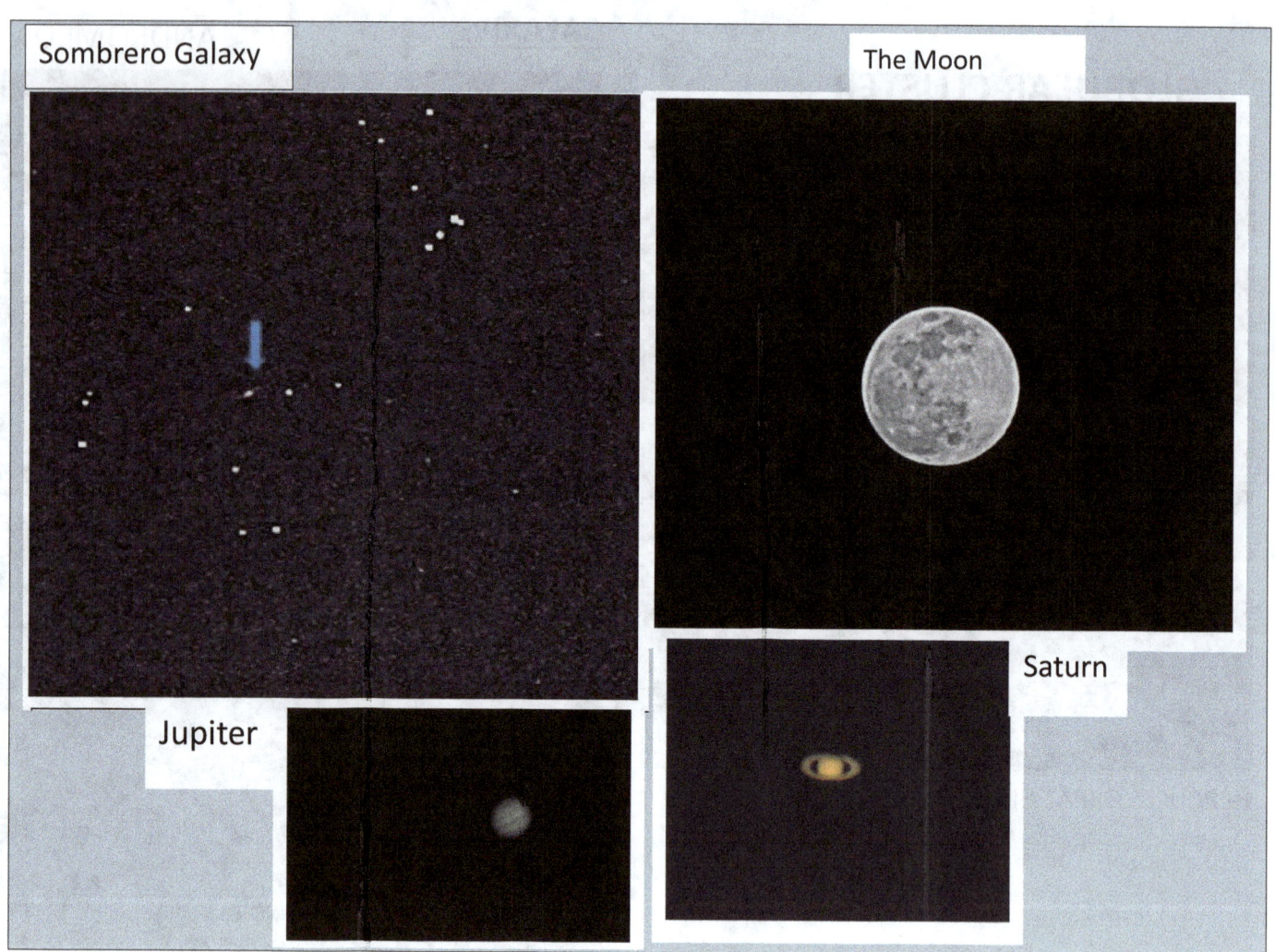

Sombrero Galaxy

The Moon

Jupiter

Saturn

NEBULAS

COMET

NEOWISE 2020

ORION
NEBULA

RING
NEBULA

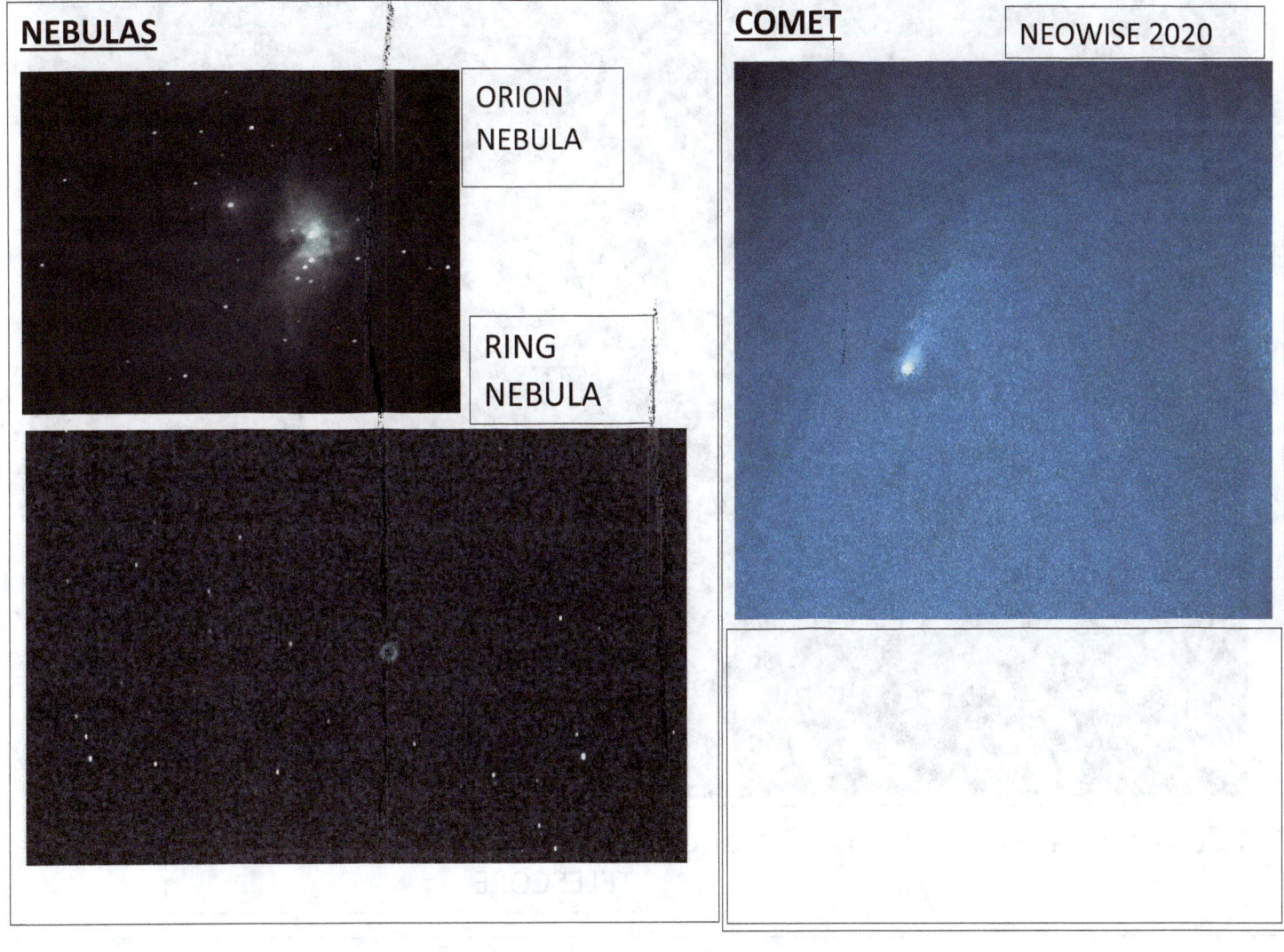

Cassiopeia, Perseus, and Pleiades WITH A DSLR CAMERA AND NO TELESCOPE

CREATION OF THOUGHTS: THE GRID, UNIVERSE, CIVILIZATION

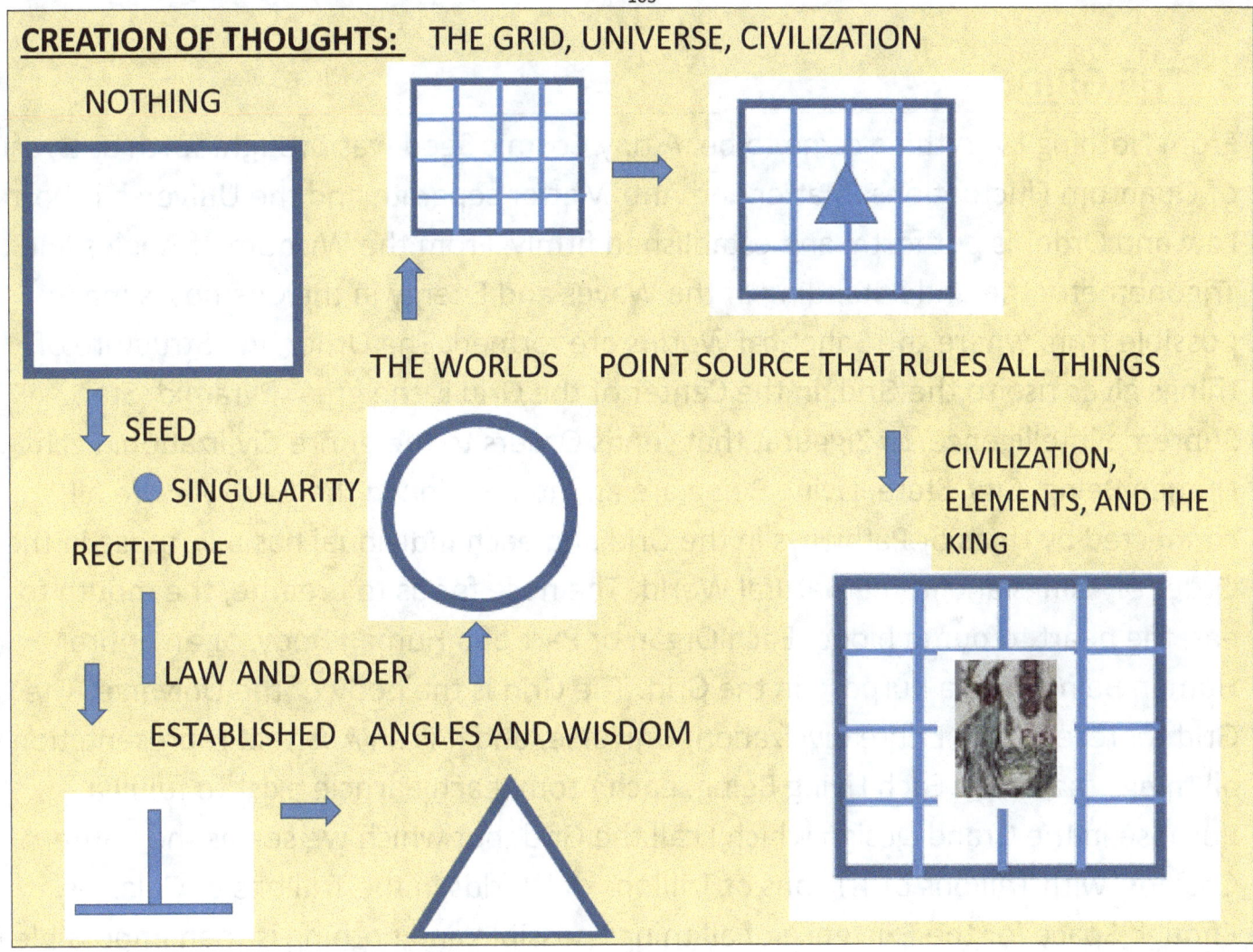

NOTHING

SEED

● SINGULARITY

RECTITUDE

LAW AND ORDER

ESTABLISHED   ANGLES AND WISDOM

THE WORLDS   POINT SOURCE THAT RULES ALL THINGS

CIVILIZATION, ELEMENTS, AND THE KING

THE GRID

From Nothing Everything came to be. A tiny Cosmic Seed was brought forth as a result of Quantum Fluctuations. Matter and Anti-Matter separate and the Universe is born. Law and Order is put forth, and established firmly. From the Wisdom of Angles and Trigonometry the understanding of the Waves and Energy in the Cosmos is made possible from where the Spherical Worlds are formed. The Order and Structure of things gives rise to the Grid. In the Center of the Grid is the great Pyramid, Star, Supreme Intelligence, or Ziggurat that sends Orders to the entire Civilization. Each Human Being, City, State, Living Being are an Intersection in the Grid. We are all connected by Lines or Pathways in the Grid and each Individual has a Purpose in the Order of Things and in the Natural World. The nose for us to breathe, the mouth to eat, the heart to pump blood. Each Organ or Part of a Human Body, or an entire Human Being have a Purpose in the Grid. The Grid is the body of the Universe. The Grid represents the Entire Civilization, Universe, and it is an Abstract representation of all things that exist. Each Living Being, each Atom, each Particle exist to fulfill a Purpose in the Grand Design which I call the Grid, but which we see as the Entire Cosmos, with Trillions of Trillions of Trillions of Worlds in the Trillions of Galaxies. Enough Space for the Existence of all things possible and nothing is then impossible in this Grand Design.

The Order in the Energy Levels of Electrons in Atoms:

Energy Level

1	2	3	4	5	6

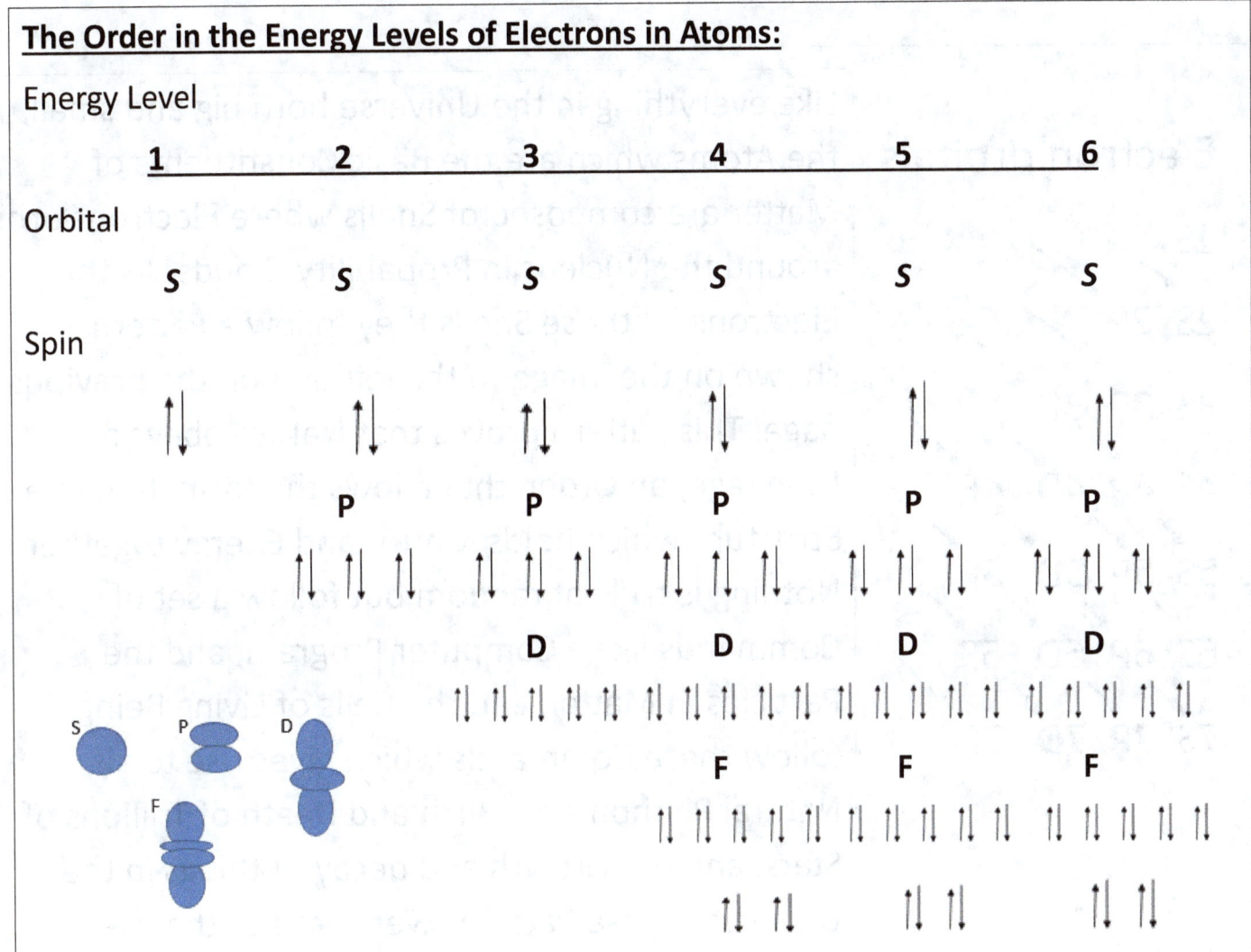

Electron orbitals:

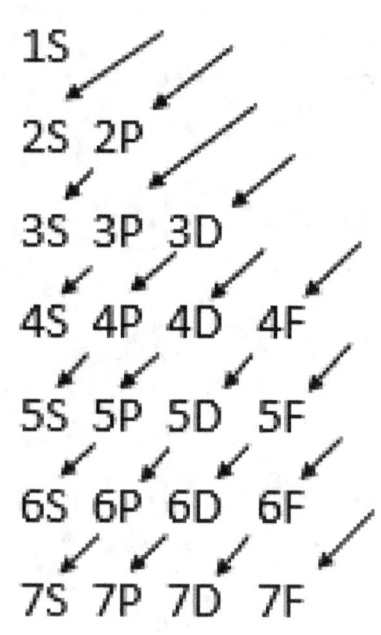

1S

2S 2P

3S 3P 3D

4S 4P 4D 4F

5S 5P 5D 5F

6S 6P 6D 6F

7S 7P 7D 7F

Like everything in the Universe both big and small, the Atoms which are the Basic Constituents of Matter are composed of Shells where Electrons orbit around the Nucleus in Probability Clouds. As the Electrons fill those Shells they follow a Pattern shown on the image to the left and on the previous page. This Pattern proves that Nature obeys a Language, an Order, that allows the formation of a Structure which holds Matter and Energy together. Nothing is truly at random but follow a set of Commands like a Computer Program, and the Particles in Matter, and the Cells of Living Beings follow these Commands which gives rise to all Natural Phenomena, Birth and Death of Trillions of Stars, and the Growth and decay of things in the Universe. These Patterns were keys to the later Development of Life.

THE WHEELS

Week Wheel

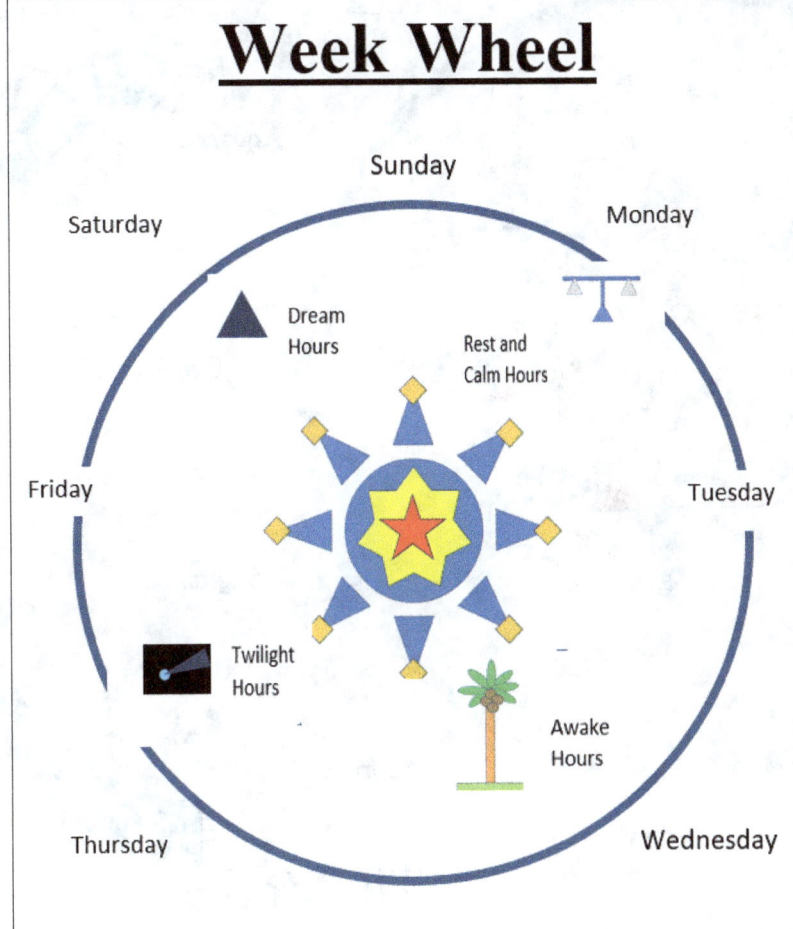

Sunday

Saturday

Monday

Dream Hours

Rest and Calm Hours

Friday

Tuesday

Twilight Hours

Awake Hours

Thursday

Wednesday

Cycles in Nature:

Nature is filled with Cycles which lead to its multiple Patterns. For every Pattern there is repetition. These Cycles are what forms the Fractal Geometry of Matter and Energy throughout the Cosmos. An Infinite Cyclical Pattern from the Macrocosm to the Microcosm. On the left is the Week Wheel and on the next page is the Circle with the 12 Constellations of the Zodiac. In the same way that there are 12 Constellations there are also 12 Months in a Year. The Heavens rotate being an Immense Clock. We Breathe in and Breathe out, Blood flows, Day followed by Night followed by Day. Cycles are everywhere.

Year Wheel

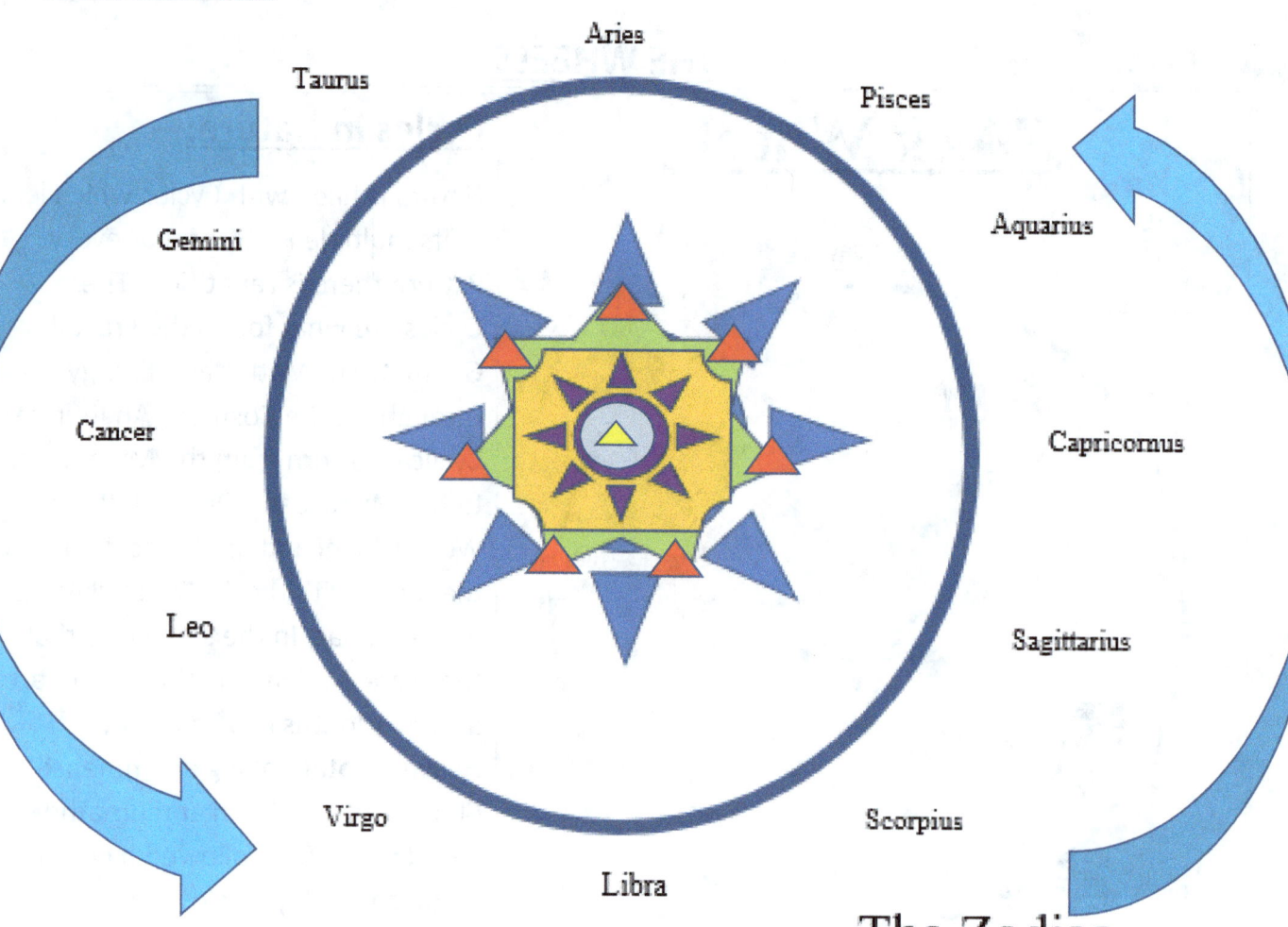

The arrows show the direction of the Sun's apparent motion in the sky through the Zodiac.

ETERNAL WEB

The bigger the Telescope placed in orbit of the Earth the more Galaxies are discovered and this becomes a reason of frustration among Astronomers. The human desire to comprehend all the Laws and Constants in the Cosmos and to have a complete understanding of a well defined and confined Universe is made difficult when it is not made possible for us to see its wall, or its edge. How can we create a well confined Theory of Cosmology if we can't the see the end of the Universe or its edge? Every time that a bigger and more powerful Telescope is built more Stars and Galaxies are found proving that there is a lot more Mass and Energy than we once thought. It does not stop there, and then humans design an even bigger Telescope to discover more of these objects placing in jeopardy all the Theories we currently have. Will we ever build a Telescope that will reach the limit, and see the edge of the Cosmos, or is it simply not possible? What if Space is truly infinite? Do we then live in an Eternal Webb of Galaxies throughout Infinity with no beginning and no end and that extends forever? I have proposed many Theories of Big Bang, growth and Evolution of the Cosmos but why not place a thought on Infinity? What if there was no Big Bang and everything is just settled in Space, not expanding, and the Doppler Effect of Light seen from Galaxies are not because they are moving away from us but because light struggles to go through dust and Gravity causing the Red Shift which led us to believe in the Expansion of the Universe? Could it be that Dark Matter causes the Red Shift of Galaxies? If Galaxies were moving away from us, were they not supposed to be looking smaller and smaller over time? Why is it that their size does not change through the Telescope? We only believe on the expansion by the Doppler Effect, but we have not truly seen the expansion itself. We have just seen the Red Shift and that is it. Do we have a complete understanding of Relativity and Quantum Mechanics that answers all of these questions? How can we be certain that there was a Big Bang if not even Einstein believed on it? Is Inflation Theory very convincing among Scientists? Not really yet!

Harmony, Rhythm, and Waves:

When writing an Essay the writer is very careful in order for the words used in the literature to flow and to follow a Rhythm that is near perfect in transmitting to the reader the Content or Topic in the way that it is intended and convincing. Everything in the Universe follows a Rhythm and a Pattern in the Flux of Events and Natural Phenomena. From Particles and Probability Clouds, Metabolism in Living Beings, to Planetary Geology, Solar Systems, and Galaxies. All things in Nature follow a Rhythm and a Pattern as seen in Fractal Geometry. Modern Theories in Cosmology, however, clearly reveals Scientists to be lost and with no Rhythm or evident Pattern in their ideas. Cosmology has not discovered a True and very Convincing Theory that can be completely verified through observations with no exceptions. The Big Bang Theory, Inflation String Theory, has led Scientist to forcibly use their imagination and write more about Philosophy and less about Science. I confess that even I have done this but where is now Physics leading to? Cosmology is lost today in the realm of Philosophy and Superstition. We create Theories to match the observations, but then we keep adjusting the theory several times in order to better fit more observations, to the point where we change them so much that we really do not know anything at that point. Too much adjusting on my opinion. Maybe the amount of Philosophical and Mathematical argument in Cosmology, Quantum, and Relativity will indeed lead us to the so dreamed Confined view of the Universe but it might take time. Until that happen we are like the Ancients creating Myths unable to comprehend the truth. This is us right now. I do believe, however, that something good about this wandering will come out of it. Although I criticize many ideas, I also know that Freedom of Thought bears fruits.

Education:

In the modern world Students of Physics and Mathematics face issues in College for an apparent lack of explanations in books about the Proofs for Theories and Equations, and questions at the end of chapters that are not so related with the content covered in these sections of the book. It is a serious problem and as the level of the course becomes more complex, an in-depth explanation is almost never provided. It is fun to read a Physics book with many Equations and Complex Concepts, but these books do not seem to provide enough information. These books do a horrible job at stimulating Critical Thinking. This lack of in-depth study and immediately assumption that students will discover in a week in order to complete a homework what took decades for Physicists to discover themselves is overloading students with work with very little expectation. Modern education expects a student to **re-discover** the Theory of Relativity alone. That is not how it should be. This shows a deficiency in the Education System, where many students lose motivation and quit. When reading several books on Higher Level Physics some of the Proofs for many of the Equations are almost never found in any of them. How can we add to the knowledge if we don't understand current knowledge without a clue on how or why current Theories work? It is wrong to just memorize Equations and do the test with no interest on why we are learning this.

Also when students are reading these books they tend to struggle also to grasp the concept or Main Idea of the Subject. They leave the class knowing less than when they entered it. It is very vague and apparently impossible. The Teachers in College also lack on their ability to fill the gaps that the Text Books skip or omit assuming everyone knows the content already. Many of these Students are seeing these Concepts for their first time. The amount of Problems that Teachers don't even cover in class, and the amount of Chapters near the end that the Course don't even cover is also very scary. Sometimes as little as 5% of an entire Text Book is actually covered in class. Why then not make a book with only the 5 %, with many Questions? Is it because not even the Teachers know what or how to teach? There is a deficiency in education, very little in-depth study on important subjects, and Students having to re-discover everything trying to understand the books, like translating Egyptian Hieroglyphics in order to survive.

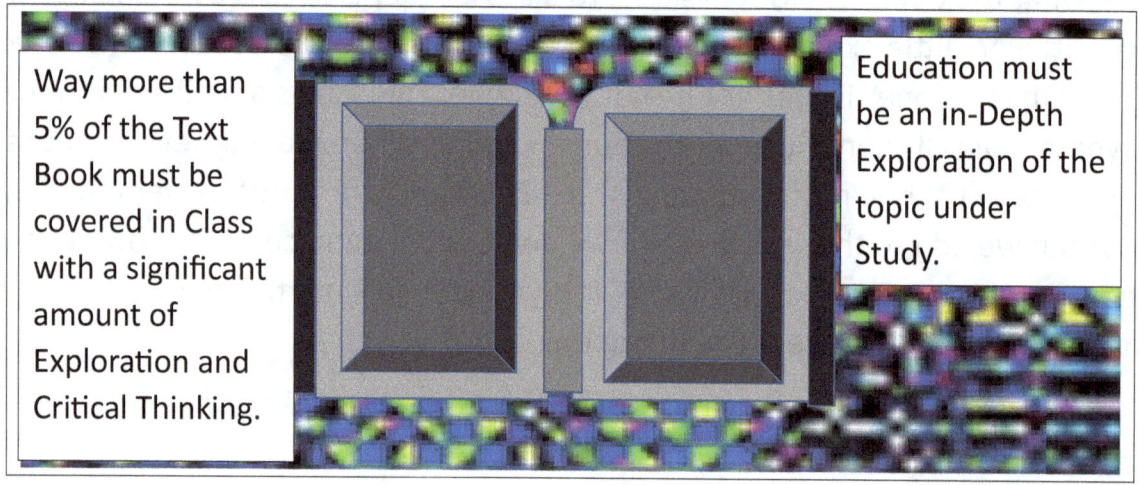

Way more than 5% of the Text Book must be covered in Class with a significant amount of Exploration and Critical Thinking.

Education must be an in-Depth Exploration of the topic under Study.

Dissecting old Physics and Mathematics Textbooks and Re-Discovering Knowledge:

The above statement which is the title for this page should make perfect sense for any College Student pursuing a Mathematics or Physics Degree. Spending 5 hours trying to decipher only 3 pages of a Highly Advanced Physics Book with so many Equations and Content worth investigating. The constant thought that you need to learn it since the test is only 2 weeks into the future. When you finally learn the Content you ask yourself why did the book not explained like this or like that, or why did the Teacher never said like this or like that. After you learn how it works, after dissecting the book, you decided to write it in your own notes with your own hints and tips, everything the way that would have helped you learn it faster without having to spend 5 hours on it. You then share your notes with other students, and your fame grows since you re-discovered knowledge like a detective searching everywhere, specially online. That is why despite of the fact that the book does not have enough Proofs and Explanations, it can still be a great Source, on your path towards a more complete understanding of the Content. I can imagine generations of the future dissecting Technology wondering how humans discovered how to build these devices centuries ago. In the same way Europeans dissected the works of Aristotle and Plato for centuries wondering how these Philosophers and Scientists developed their Theories. Dissections are fairly common and we as a civilization will only make progress if more of these dissections are done in the future. Humans from the future trying to understand us in their past.

Education: The Learning Cycle:

The Greek Philosophers were all Teachers of Life, Science, and many other things. What made the Greek Civilization to prosper in its Scientific Pursuit was its Philosophers drive for Teaching, Observing, and Learning. We are currently living in an age where Technology is common all around, but people know very little about it. The Modern Civilization more than ever needs Teachers. In the same way that the Ancient World would be completely poor in Knowledge without the Greeks, today the world is delving in ignorance because it needs Teachers more than ever. Our youth is very lost, depressed, and has little motivation. There seems to be many mental and emotional disabilities, difficulties with learning, social problems, and a feeling of loneliness. People blame the Technology, the Social Media for all this mess. I used to have Social Media but I do not anymore. I use computers to write, publish my work Online, and to Teach Students. There are ways for the Technology to actually be a benefit when it comes to Education. The Internet is one the greatest of all Human Inventions. We must thank the United States of America for giving the entire world access to this limitless source of Information and Knowledge. Babylon has its Tower, Egypt its Pyramids, Greece its Philosophers, Rome its Warriors, France and Germany its Mathematicians, England its Industrial Revolution, and America the Internet.

The Internet is still a brand new Invention and I do not think humans have truly learned how to use it properly yet. The Internet which was meant to Instruct and deliver Information, also became a source of false information and a delay in the learning capabilities of our youth who get too distracted easily. Brand new inventions take time to become a common a thing, and the process for that to happen involve creating laws to strict Internet usage. These laws would prevent people to publish certain material that is offensive, detrimental to human soul, or false. Everything posted online would be filtered. The wrong or offensive picture or information would be deleted. There would also be laws in Social Media preventing posts labeled as not healthy for humans, or offensive, or dangerous. The Internet is a Virtual Reality and Virtual Universe parallel to our Reality. In the same way we have Armies, Soldiers, and Police in our world, we need protection for Internet Usage. Over time humans will create a way to improve the Internet only allowing Information and Content that is healthy for the Human Civilization. It will still allow plenty amount of Freedom of Speech as long as it fits within the Lines of Prosperity, Justice, Respect for Human and Animal Life, and Ideas that can be beneficial for the World in General. The youth is in trouble today because they are having access to worthless Content Online. The Usage of the Internet is a Key Tool to actually Improve Learning Skills and lead to Progress in our Civilization. One day we will know how to that more efficiently.

The Learning Cycle

The first step into learning something new is by starting an Inquiry such as Asking a Question related to the Topic under investigation. Science is about asking questions and deriving an explanation for the Event supported by facts as seen in the observations. After asking a question, a Scientist must make an Educated Guess supported by previous research or experience. This guess is known as a Hypothesis which may or may not be correct. Next it is necessary to conduct an experiment that includes all the factors involved in the question and must also include testing the hypothesis in order to confirm or negate it. The Experiment will need Technology, some recording device to store data, and to take notes on observations. Data must be collected with multiple trials, and notes must be taken on what was used to avoid errors, and an explanation for all the carefulness used and safety protocols to permit a safe and free of bias result. Data is then analyzed with the use of Equations, Line of Best Fit, and Uncertainties are calculated and compared with the average of the results. This investigation allows the researcher to know how uncertain are the results and how valid the Experiment was into answering the question. The Line of Best Fit then allows predictions, and may lead to more questions involving Proportionality, Trends, and a New Hypothesis that will require further research and experiment. The Results will be used to confirm, or negate the Original Hypothesis, or adjust the Hypothesis to be more consistent with the data. That is why more experiments are necessary and this is a long unending process since discoveries happen everyday.

At the end Scientists create a Statement based on a Conclusion that can be made possible with the Data Collected, and these statements may one day become Theories with Equations, Laws, and Constants. That is how the Scientific Method works and it is how Students should learn in a classroom. The Learning Process in a Classroom should be the same process that Scientists use in developing their Theories and verifying their Hypothesis. It is important for Educators to encourage Critical Thinking in class, allowing Students the opportunity to experience being a Scientist, by following the Scientific Method of Reasoning, the Inquiry Based Process taught by Ancient Greek Philosophers.

LOGOS

Learning comes from Observing.

Learning also comes from doing it.

<u>Sumeria:</u> A great way to live Philosophy and to have a more complete understanding of the Learning Cycle is also through observing ourselves. Humans began living in cities 6000 years ago, and with the constructions of these settlements came Art. Unlike other animals in Nature, Human Constructions, buildings, roads, towers, and so forth are always closely related to Art, with Colors, Patterns, Texture, Fractal Geometry, and these reveal Human tendency to place meaning in all they do and learn.

The City:

The passion that Humans have for patterns and texture can be found in everything we do. Not only in the Arts, but when organizing a home or a room, writing a book, cooking, playing sports. All of these follow a given set of patterns and rules in order for it to work. When we observe the Natural World we see the exact same thing. The Seasons of Year is a pattern, the motion of Celestial Objects obey cycles, the breathing of animals repeats over and again, the Cardiovascular System functions with the Heart pumping blood repetitively, the Cells' Structure and its Metabolism, the DNA Codes leading to Genes and Traits, the petals of a Flower also obeys a pattern, the Spiral Structure of Galaxies in the shape of a disc, and it goes on forever. When designing a Lesson with the attempt to instruct students on some new topic, it is important to first ask them a question and guide the lesson by asking further questions until the student, based on previous known content or past experience, starts answering these questions right. It is the purpose of the teacher to give the students opportunity to realize the patterns of nature, delving deeper into the essence of all things, the repetitions, the cycles, how one topic in Science relates to other topics, how all things are very connected to each other. The student is guided though an Inquiry-Based Process while being introduced to the flow of thoughts and being driven to perceive these patterns, with the Critical Thinking being constructed towards a more complete and confined understanding of Science. The Concept of a City clearly shows interconnectedness. How each individual contributes for the City to grow, and prosper. The flow of income between business, banks, and houses, the Police guaranteeing safety, the Laws and Rights found in the Constitution, the preserving of the City by keeping it clean and safe also reminds us of how organized are the Cells in Living Organisms. In the same way that Nature shows a high level of organization, Human Cities, Computers, and Robots also obeys a clear set of Patterns or Codes which allows just about anything we do, or that we see in the Natural World work properly.

The Nation: A nation is composed of a population of people with a culture bond together by a language and moral values. There are Cities and States spread throughout the Land represented in the image below by the Thick Towers with a red star at its top. The Radios, Cell Phones, or any Wireless Communication are transferred with the High Thin Blue Towers while the Internet Servers are the Pyramids. In our Modern Age we can see how Technology connects all the Nation and all Countries in the World, allowing Fast Communication which facilitate our understanding of each other and our access to information.

The Internet and Neuralink:

A single butterfly flapping its wings in the Amazon Rainforest causes a disturbance in air that keeps adding changes to the flux of Molecules. This leads over time among the Chaos of Random Motion, to events that would be different if the flipping of the wings were at a different Angle, or if there was not butterfly at all. Everything is deeply connected to each other. One decision in a faraway country will immediately change the course of History and affect an individual on the other side of the Globe. The Natural World is all connected and the Human World is now more than ever through the Internet. One thought I have today, and write in a book, a year later I publish Online and people can read it. One idea of mine posted in Social Media and may lead to other people thinking the same way, or arguing with me. People in the farthest country can read my posts in a single second, and my thoughts has been transmitted across Continents. This is Humanity today fully connected Online. If I have a question I can ask Google. If want to chat, I can text people and post Videos Online and later read people's reaction to it. The Internet is one of the biggest inventions of Mankind and will be a key tool in the development of Artificial Intelligence. Imagine if Google becomes a robot, a counselor, a creature that will have the answers to all your questions, and a great object to ask for advices in life. If somehow these databases that allows the Internet to function were all downloaded in a Hard Drive of a Robot, we would have built the smartest object on Earth. A Super Intelligent Robot that knows everything. I can see humanity being ruled by Super Computers in the future and Earthlings asking these machines questions on how to solve major world problems. Another way that the Internet can revolutionize Human Civilization is if somehow we can wear a Helmet and be transported to a Matrix, a complete Virtual Reality.

These Virtual Worlds that Humans would be able to venture could be used to teach us about Science and the games played would be simulations to train soldiers for war, and possibilities that are numberless. Imagine studying about Mars but not being able to go there in person. We could just put a Helmet over our heads that would transport us to a Virtual Mars that looks identical to the original. Students would then be able to study about the Planet without having to endure a long trip and a dangerous exploration of the Red Planet. The Internet is the future, and we are yet to see its full potential.

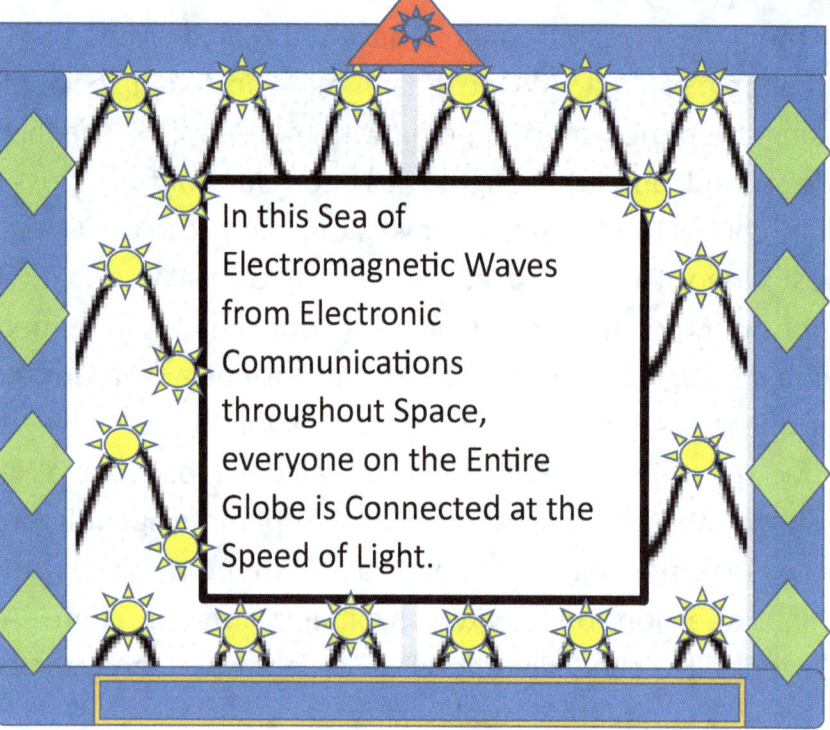

In this Sea of Electromagnetic Waves from Electronic Communications throughout Space, everyone on the Entire Globe is Connected at the Speed of Light.

Artificial Intelligence

Substratum:

Computers will over time teach humans everything about ourselves and the Universe. The Internet allows a rapid flow of information that has the Potential to lead Humanity into a completely different world in the future, where countries and borders will be no more, and where everyone will identify as belonging to a Planet instead of a specific region within the Planet. Each person will identify as a Human, without specifying location, culture, or language. The Internet will provide the way for total Globalization of Ideas, and where the world countries will come together to lead the civilization into solving all of the social, health, and economic problems. This is called a Planet Civilization filled with Technology, and where every person is connected to each other Online. Education is extremely important and that is why Teachers must never lower the amount of Content taught in Schools but should instead discover a way to Elaborate and Enrich the Lessons that will be more appropriate for current Students without making the course any easier. Students today require a different approach to Teaching that must involve Technology and the lessons and activities should be creative enough to captivate their attention. We learned from Ancient Greek Philosophers the importance of dialogue and Critical Thinking through and Inquiry-Based Learning. The Philosophy of Ancient Greece must be taught in Schools since the Ideas at that time was a turning point for the Revolution in Europe after its Dark Ages. The Greeks were the first to formulate the Scientific Method of Reasoning, and they mark the rise of Western Civilization's view of Ethics, Culture, and they established the Foundation for Modern Politics and Democracy. Every person is an important Element of a City, Nation, or State. Every Element has a Purpose in the Grid, in the same way that every part of a Human Body is equally important for Life and Survival. All things have a Purpose, and nothing is for nor reason.

Aristotle spend much of his thinking if not all of his thinking wanting to understand about the Nature of All Things and there is among his works a profound study of the Idea of Substance. Take for example Cooper as a Substance from which it can be molded into a statue of a man or a cat. Both a man and a cat are different, but being both statues of Copper they both have the same Substance. All things are derived from something else. All Matter came from Energy, and Energy could very well come from Matter. From Energy or Light a Particle and an Anti-Particle are formed, but once they meet again they turn into Energy in the Form of Light. All Particles have Energy, and all Energy is a form of Light. Let there be Light and all of Creation was brought Forth. Life also has to do with Water. In the Waters the Spirit of Life was settled and the First Cells were formed. There are Primary Substances such as Particles of Light, and there are Substances that comes from them such as Atoms, and which forms Matter, and so forth. The Primary Substances are the first in a series of steps and combinations leading to all the diverse materials we see in the Universe. Everything is like Lego Pieces combined together forming Molecules, Life, and materials in all Four States of Matter: GAS, LIQUID, SOLID, AND PLASMA. We live in a reality of multiple Chemical Reactions and our bodies are a Biological Machine with a great amount complexity almost like an entire Universe inside everything alive. It is truly an amazing work of Art, is Nature with all of its details. No wonder the greatest of all the Arts that people have created are rich in detail. Anything as detailed as the Natural World is worth praising. Nature has a significant amount of Content, which is why books rich in Content are also praised. Nature is amazing!

Categories:

The Universe is very chaotic. Particles keep bouncing off each other, and Atoms combine or break apart in huge explosions. Stars explode, Stars shine, the Galaxies spin, Planets get hit by Meteors, Particles move back and forward in time. The Universe is a Space filled with events, and here on Earth we experience moments of danger, sadness, other moments of joy, party, and our emotions run following the flow of these events. Among the multiplicity of these historical events whether at the macro scale or micro scale, humans tend to label things, in the same way that we label the Stars, and give names to Constellations, we categorize all things in our language, in our world, in everywhere. We give names to pets, as soon as a person is born we give him or her a name, we are constantly labeling things, whether they are Particles, or Hurricanes, or mountains, rivers, and even features found in other Planets. There is not a single crater on the Moon without a name. We categorize things and people and we create hierarchies. We call a person a king, another we call a prince, others are scribes or Scientists. In a house, we label the kitchen, the bedroom, the living room. This is a sign of intelligence when things are broken apart, labeled, and clearly defined what is and what it is not. We compare and contrast two things or multiple things. There is a vast Universe to explore, and even the Blind Nature which moves and follows determined Laws and Constants might not label things through words since it is Blind, but acts in such a manner as seen in Natural Phenomena that leads us to Categorize its multiple parts into generating a Whole Picture which we call the Entire Cosmos a Fun Place to be.

Emotions, Math, and Hormones:

Everything follows a Pattern. At both the Macro and Micro Scale, Nature is a Fractal Geometry. There are a series of repetitions and an infinite amount of possible shapes and structures that holds Matter in place. Our Mind is a Thinking Biological Machine filled with mysteries that are worth investigating to solve many of our problems including emotional. What leads to a personality are the factors found inside and outside of us. In a very good and pleasant environment, everyone will be inspired to do things well and within the flux of the minor and major events around that contributes to even more joy and peace. That is because the Mathematical Description of these Waves in the air where the person is located are very optimistic and filled with creativity, joy, and peace. Emotions are Mathematical such as the distinction between being Very Happy, Just Happy, Neutral, Not Happy, and Very Sad. All things are Waves and are Mathematical especially inside our Emotions. The amount or type of a specific Substance that a Hormone releases which leads to moments of joy, peace, and satisfaction, or the opposite. Math is everywhere even in how we feel. When placed in a very negative pessimistic environment, we absorb these horrible feelings like a sponge and that leads us to also join that flux of events that could move us to a feeling of hell and suffering, and all destructive types of things. That is why when we listen to Music, the vibration of these Sounds could either inspire you to something good, or something disastrous. Same applies to what is watched on Television, or what books are being read. We are strongly influence by these many factors and there is a need to be in the right place at the right time, and we need to always do what is right so that we will be safe and surrounded by the good vibrations.

Music:

Music is composed of Sounds which are a vibration in a Gas, a Liquid, or a Solid. Sound requires a medium to propagate, and in most cases, we hear Sound through Air. This disturbance propagates from the Source of Sound into our ears. The vibration in our ears converts Sound Vibrations in Air into Electrical Signals that travel to our brains allowing us to hear. Sound has the potential to lead our thinking, driving our emotions, guiding our thoughts, reminding us of the past, giving us hope of future events. Music has a large impact in our lives. The Music from different Cultures allows us to feel and experience that Culture more than words can ever describe. Music is a language that involves emotion, and all Songs have a secret message among these Vibrations that leads the listeners to that message. Our Brain is like a machine and is capable to translate Sounds into feelings. This is why Music can be both good or bad for its listeners. What you hear over and over again could cause you to make a mistake, or if the Frequencies are good, and you are driven towards something good. The entire Universe is a Hilbert Space, a Sea of Quantum Fluctuations, that gives shape to Matter and Energy. The entire Universe is a Vibration with many details and Frequencies that add together into generating our Reality. There are good and bad Frequencies and we fight a daily battle into surviving in the Sea of good, but sometimes controversial Source of Vibrations. It is important to only listen to Sounds that makes you good, optimistic, and healthy. Most people, however, specially the young, listen to detrimental Frequencies that causes depression and a feeling of failure in life.

Justice:

In the same way that we must only listen to Music with Sound Vibrations that will lead our mind to translate these vibrations into thoughts towards good feelings, and peace, the Society as whole needs a good Concept of Justice for there to be Harmony among us. The entire Universe is a vibration and the Human Civilization can only prosper if it is ruled through Justice rather than by Force or Power. When a Nation is ruled by the Powerful and Greedy, its people can betray each other, in an environment filled with Hatred and Pride that is spread everywhere, leading to Wars, diverging ideas, vandalism, rebellion, and eventually its collapse. That is how the Roman Empire fell, and that is how all powerful kingdoms in history reached their doom. In order for a Nation to remain standing, strong, and prosperous, it must be ruled by Justice. In a just system of Government, betrayals are oppressed, people are taught to accept each other, everyone has a clear and common goal in mind, and all states, people, and cities are bond into an agreement, all bowing to the same flag and Culture. The most corrupt and unjust Nations are the poorest. The Countries whose ideology is Freedom, Democracy, Justice, and Peace are able to prosper economically, eliminate corruption, and remain standing as long as these values are kept in high regards. Justice is the only way an Empire or Nation can reach its full Potential and even conquer the Entire World not only in territory but also by its Science, Ideas, Philosophy, and Culture, leaving a Legacy for many Ages into the Future. It is Justice rather than Power that is the Key for any Success.

Mysteries, Unknown Phenomena, and UFOs:

Throughout history humans have created myths, believed in fairy tales, feared the unknown, and Modern Age could not be any different. Truly in the Universe if we begin looking for so many things for so many times we will in fact eventually find things that we can't explain. The more Scientists search in the Natural World through observations, collecting data with the usage of Technology, and performing Experiments, new findings then forces the update or creation of new Theories in order to explain the unknown. Science is a constant quest and attempts into explaining the new findings with new Theories. The letters UFO stand for Unidentified Flying Objects. These are objects found flying in the air which Scientists have no clear explanation. They could be spies from other Countries, balloons, meteors, or something else which we do not know what it is. It is fun to imagine that they could even be spaceships of time travelers from the Future visiting the past which is our current present. Others think that they could even be Extraterrestrials Beings, coming from a different Planet and visiting ours and causing overall panic. Imagination truly is bigger than knowledge but we must always focus our attention to what is more obvious. If there is a sound over the roof, most likely it is not a ghost but rather something physical such as leaves from trees, or birds, but not ghosts. In the same manner an UFO should not immediately lead people to believe that they are Aliens, but we should rather think on possibilities that are more physically obvious. Only after we eliminate all known Physical causes, and if despite of that the mystery can't be solved, then we can say that we do not know what they are, and we should avoid referring to myths or fairy tales. There is also that question whether if time travel is possible, then could we not say that these UFOs are from our future?

Vanity, Pride, and the Power of Simple:

There are so many mysteries in the Universe and unexplained events that happens everyday many of which we don't see directly. The Universe is a very large place and even possibly be Infinite in a Forever Sea of Quantum Fluctuations. Humility is then very important in life in face of all this diversity and colossal structure that creates the reality in which we live. In this Cosmos despite being complex, overthinking in an attempt to understand it led many to trouble. Things can truly be difficult but in order to understand these complexities we must observe and record the facts and simplify the fractions. It is by making difficult things easier that Scientists are able to explain the entire Natural Phenomena with a single Equation. The belief that reality is too difficult and impossible to describe with words or numbers is false and I classify it as overthinking. Now formulating an Equation and stating a Theory with Clear Statements, is thinking correctly by simplifying the complexities of Reality into Discrete Statements, Laws, Constants, and Equations that allow Humans to make predictions and have a well-defined a confined understanding of the Universe. It is through simplicity rather than vague words that we can reach the Ultimate Understanding of all that there is. Humility then is the key for success, since pride is overthinking and filled with alienation. It is by making things easy rather than more complicated the way Critical Thinking should work.

A Circle or a Wave:

As explained earlier, the greatest of all teachings about Matter and Energy is the fact that everything in the Universe can be described with a Circle (Cycles) or with a wave (Oscillations). We should engrave this on stone:

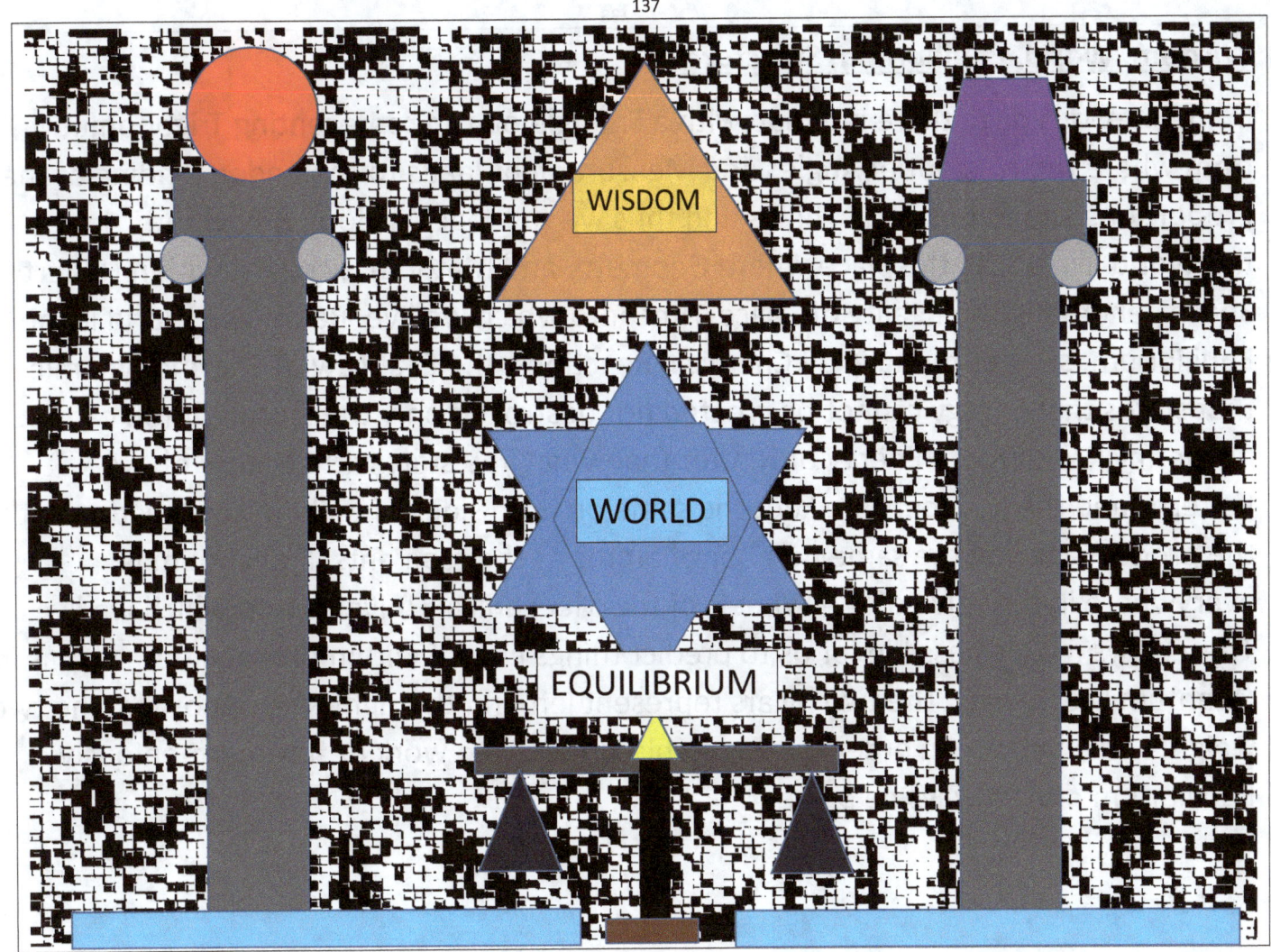

Wisdom, World, Equilibrium, Structure:

In the previous page there is a map with a Triangle on top representing Trigonometry from where the concept of Angles and Measurement leads to the understanding of the Proportions and Ratios of all Things and thus we call that Wisdom. Below the Triangle is the World like a Star, the place where Geometry and Mathematics exist within Hilbert Space generating the Reality of Space-Time into which we live. Below the World is the Equilibrium, such as found in all Chemical Reactions, the First Law of Thermodynamics that states that Energy cannot be created nor destroyed only transformed. You can never finish with more Energy or Matter than what you start with. Nature obeys Patterns and the Numbers are kept the same on both sides of a Chemical Reaction with the products having the same number of atoms as the reactants. Equilibrium is everywhere in the drive of the Wheels of the Mechanical Universe into which we live. These Laws makes possible for us to predict things even if within probabilities such as in Quantum Mechanics. The two pillars represent left and the right, the opposites, the two weights on a balance in order for this stable Universe to work or flow harmoniously Equilibrium must always be present.

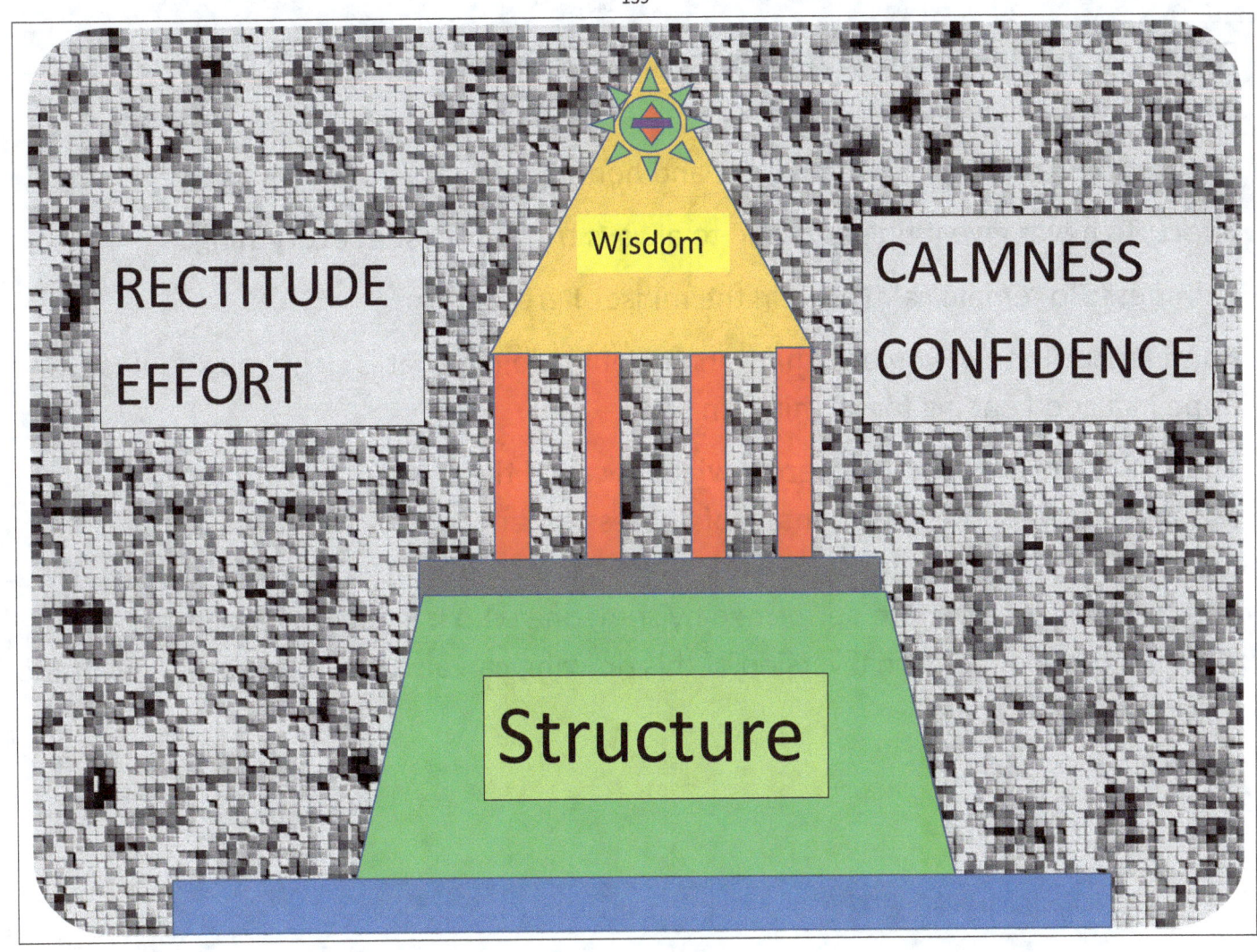

Foundations: In the previous page there is the Triangle of Wisdom being held by 4 Pillars. These Pillars stand for:

Rectitude: To always do what is right and honorable.

Effort: To never give it up easily and to always try your best at everything.

Calmness: To remain calm even in the midst of great storm.

Confidence: To trust you breathing and through doing what is right you should always hope in a good ending for all things in life.

These 4 Pillars are over a Trapezoid which is called the Great Structure for this Great Mountain of Wisdom and Fountain of Success for anyone who are always under the Law, never slothful and ready to work, Calm and free of stress, and Confident on the end result of every Action performed by someone who has the correct attitude and vision. A three dimensional version of this drawing can also serve as a great decoration for a home or office.

Learning from Artificial Intelligence:

The future of Human Civilization will indeed require Robots among us due to their Great Potential into helping us better understand our Human Nature. The development of Robots and Computers is a long process that includes addition of details, data storage, and pathways in Electronic Circuits thus increasing the diverse capabilities of these machines. There are more than a Billion Transistors in a Cell Phone Microprocessor. It is truly a work of Art, and every new version of a Phone, or Computer, is given a better and a more capable Hard Drive with a capability of processing data and performing calculations at a faster rate and also the ability to store more data. The research into Artificial Intelligence will improve our understanding on how the Human Brain works, in a process of enriching the processing of a Robot's Intelligence, improving its reaction to events such as the detection of radiation, light, sound, and how the Robot responds to these like a person does in real life. How to make a Robot be able to recognize a person and start a conversation, or giving the Robot instructions on how to avoid hitting a wall while walking, or bump into someone on the way. It is also important to download a Software in the Robot that will under a meticulous process allow the Robot to think and learn all by himself. That means that in making Robots closer to a Human, over a long period of time with research and observation, will also help our understanding about how our brain works and solve the great mysteries of the mind and our self-identification.

Space Exploration:

In the same way that designing Robots or Computers will allow us to have a greater understanding of our Human Nature, Space Exploration will indeed teach us more about Our Own Planet and the Universe and solve mysteries such as the birth of Life. Through our gazing at the immense Heavens with an apparently infinite number of worlds, it is obvious that Earth does not hold any special place in the Universe but is rather one more world among multiples. At first it would be more convenient to build Satellite Cities in orbit of Earth or Mars or some other Planet. These would be shaped like a disc that spins generating Artificial Gravity. Astronauts in these cities would perform experiments in Micro Gravity and discover ways on how to live in Space with the least amount of negative impact on individuals. We can send Humans to Mars without landing on Mars, but rather entering in orbit of the Red Planet and contemplating the Planet from a distance. These Satellite Cities would serve as great Space Stations where research and experiment would be performed all the time. After a significant amount of Study and decision on what precautions to take, Humans would then Land on Mars or on the Moon and build Stations on the ground of these Asteroids or Planets. The First Goal is for humans to live in Space in Satellite Cities in orbit of a Planet, then we can later send Humans to the Planetary Surface, and last build a Station on the ground. A Space Station on the Moon would be a great starting point of a Space Trip in order to launch Rockets since the Moon does not have an atmosphere.

Steps into Space Exploration:

1....Build Satellite Cities in orbit of Earth and Mars.

2...Send Rovers and Robots to the Moon and Mars and have them build Space Stations on the surface of these Asteroids.

3...Land people on the Moon to help in the construction of the Lunar Space Station. The Moon then becomes a perfect place for launching Rockets to Space Missions in the Solar System.

4...After studying all possible factors in the Mission, humans then land on Mars with the risks for this trip being reduced.

5...Build Satellite Cities in orbit of Venus, Jupiter, Saturn, Uranus, and Neptune.

6...Land on the Jupiter's Moon Europa.

7...Build Large Telescopes on the Moon's Surface.

8...Start Colonizing the Moon with migration of People.

9....Start Colonizing Mars with migration of People.

10...Build a Large Spaceship with hundreds of People venturing beyond the Solar System into Alpha Centauri.

Ideal University:

The perfect University teaches the Sciences from a starting point that is strongly linked to Philosophy. Ancient Science was called Natural Philosophy since it is by definition the Human Concept of the Natural World with Theories and Equations. The second part of the University is the Hospital which does not only offer treatment for the body but also a therapy for the mind, cleaning an individual's emotions. The Hospital should treat the individual both their mind and body. Last comes Gymnastics that instructs people to exercise, run, swim, lift weights, and so forth. These are the three parts of an Ideal University.

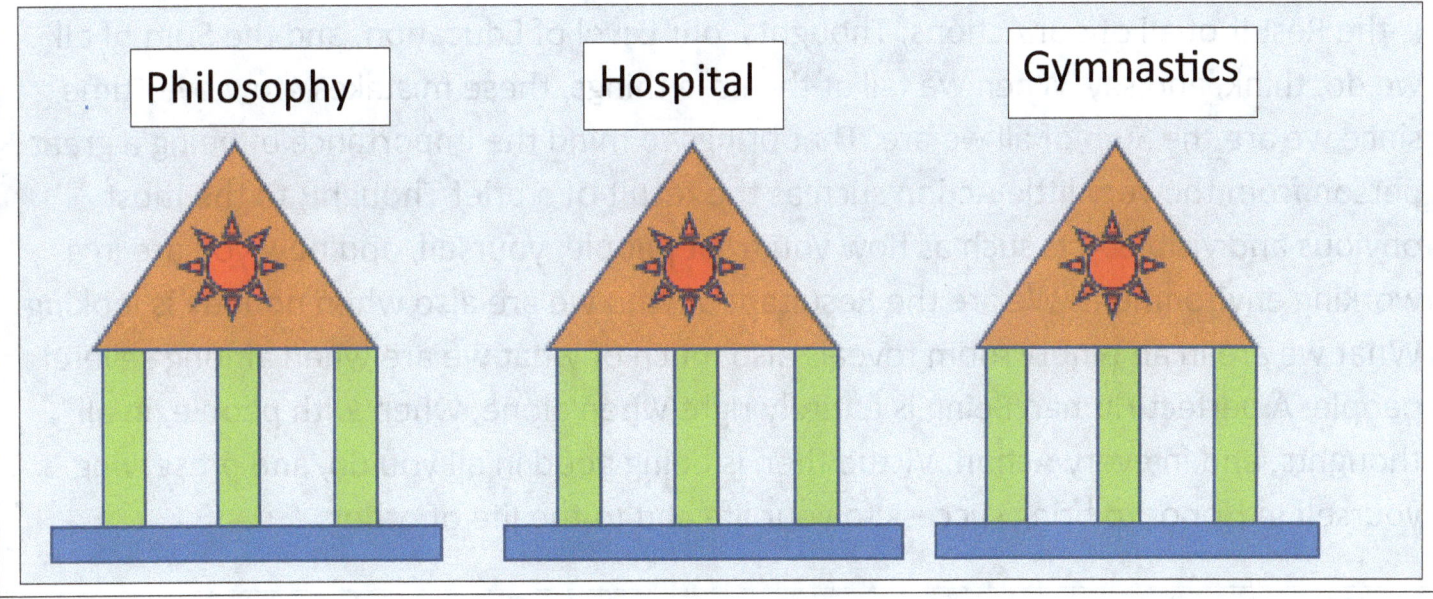

Philosophy   Hospital   Gymnastics

Courage, Honor, and what is Right:

The Flux of Events that happen in a person's life can be good or bad. As mentioned earlier, the Universe is a flux of Energies in the form of Waves giving form to Matter. When adding all of the Wave Functions such as in Fourier Series, by summing up all of the combination of Sine and Cosine Waves we get a Table, a Chair, a Computer, a Dog, a Person, and so forth. Everything that exists is the Resultant of the Sum of Waves. In the like manner, what leads to the Acceleration of an Object is not a Force but the Sum of the Forces. Likewise in our Actions we build our Reputation. Our Honor and Reputation is the Result of all of our Actions, Thoughts, our Level of Education, and the Sum of all we do, think, and say. When we fail at the little things, these mistakes grow over time since we are the Sum of all we are. That brings to mind the importance of being a great person from the very little Action such as the result of a brief Thought, to the most obvious and visible Acts such as how you treat people, yourself, and how you are in a working environment. We are the Resultant of who we are also when nobody is looking. What we are in an empty room reveals also much of what we are when among several people. A perfect Human Being is entirely right when alone, when with people, in all thoughts, and in every Action. Virtue then is being good in all you do, and preserving yourself in Honor to bring success to your life and to the life of others.

The Ideal State:

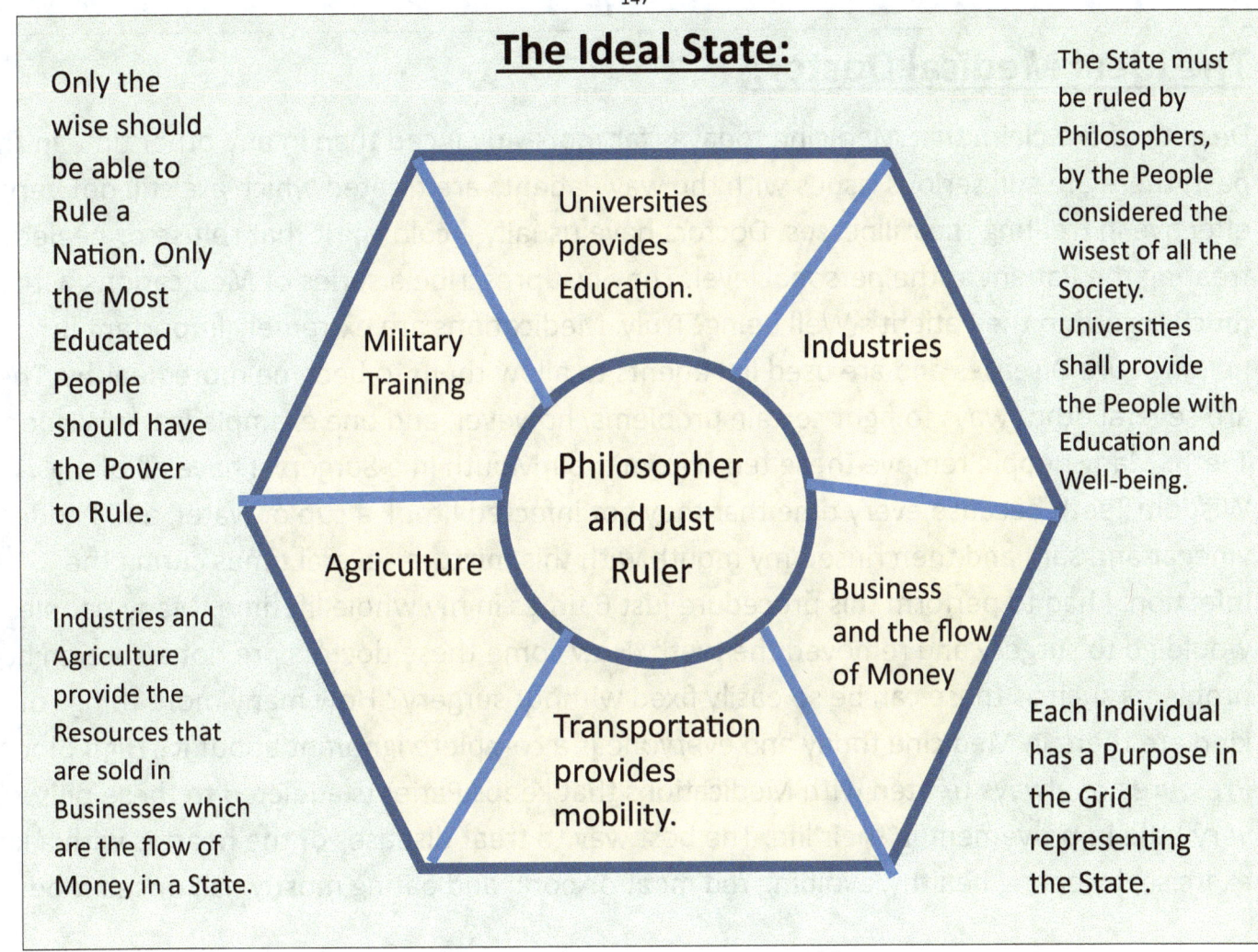

Only the wise should be able to Rule a Nation. Only the Most Educated People should have the Power to Rule.

Industries and Agriculture provide the Resources that are sold in Businesses which are the flow of Money in a State.

The State must be ruled by Philosophers, by the People considered the wisest of all the Society. Universities shall provide the People with Education and Well-being.

Each Individual has a Purpose in the Grid representing the State.

Universities provides Education.

Industries

Military Training

Philosopher and Just Ruler

Agriculture

Business and the flow of Money

Transportation provides mobility.

The Ideal Medical Doctor:

Despite of the claim that Medicine today is far more advanced than in any other time in the past, there are still serious issues with the way Patients are treated which are still not very effective in treating their illnesses. Doctors have usually a cold spirit that refuse or neglect treating the Patient at the personal level. They just prescribe a series of Medications without much regard to the Patient's Well Being. Truly, Medications are extremely important in helping cure Diseases and are used in Patients to allow them to become more healthy. There are several other ways to fight certain problems, however, and one example is the Wisdom Teeth. Many people remove these teeth from their Mouth in a Surgery. I have all of my Wisdom Teeth because every time that they got infected I took a cup of water, and I added vinegar and salt, and then rinsed my mouth with this mixture several times curing the Infection. I had to perform this procedure just 6 times in my whole lifetime. Many people would go to surgery and removed the teeth. How come these doctors are not aware on how problems such as these can be so easily fixed without surgery? How many more things of this kind are there in Medicine today and everyone is a complete ignorant about it? High Blood Pressure are always treated with Medications that keeps Patients addicted to these pills with very little improvement in their life. The best way to treat diseases of the heart is with a light exercise, by eating healthy, avoiding red meat or pork, and eating mostly fish or sea food.

The Doctors, however, prefer to fill the Patient with pills with very little understanding of the Patient at the personal level. Everything is treated with either a surgery or medication. I call that brute force with little knowledge on how a Human Body and Well-Being works. This indifference to a Patient's Health is detrimental to the Society where we now have several people addicted to pills, not knowing that a walk early in the morning, and a completely different diet would be a great substitute to those pills. In the world today, Ignorance is more common than the Absolute Truth, and in Medicine today it is the longer and less effective way of treating Patients that doctors use the most. For many other things, however, Medications are in fact important and so are Surgeries. The problem is not on the Medication or on the Surgery, but it is in believing that every disease will need a Surgery or Medication. There are several Diseases that can be more easily fixed, but overthinking of doctors, and their indifference into understanding the Patient at the Personal Level leads to more pills, and more pills, and more pills, and addiction, more depression, and very little Therapy. Medicine today is cold and is a subject still far from being able to completely understand the Human Body and even less about the Human Soul. There is still much to Discover and to improve.

Thinking is Good but there must be Carefulness from Overthinking which is a waste of time.

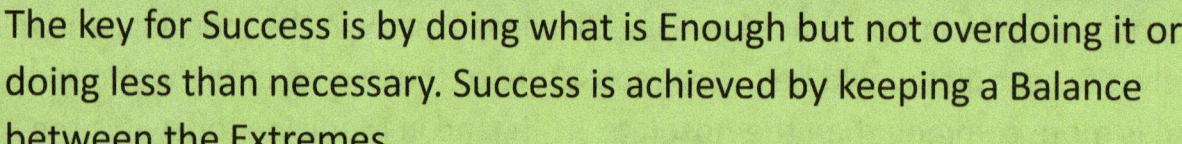

The key for Success is by doing what is Enough but not overdoing it or doing less than necessary. Success is achieved by keeping a Balance between the Extremes.

There is only one Absolute Truth and not many that describes the Natural World. All Theories must converge to a single Equation or Statement.

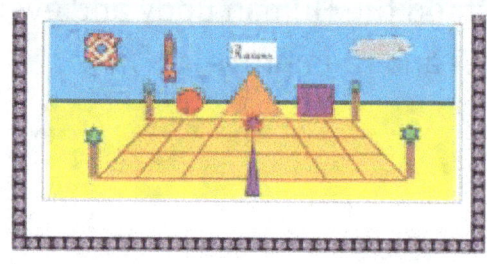

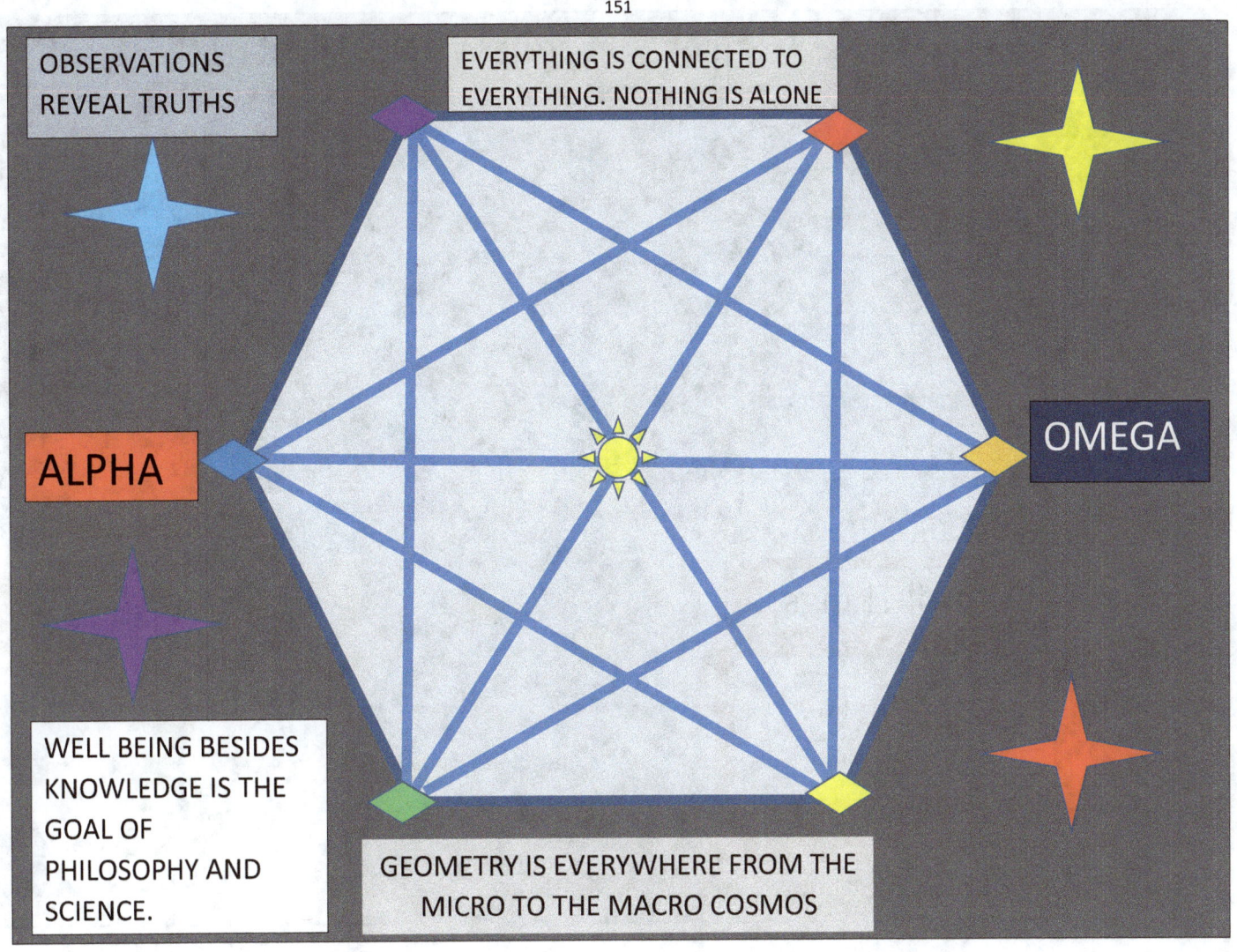

OBSERVATIONS REVEAL TRUTHS

EVERYTHING IS CONNECTED TO EVERYTHING. NOTHING IS ALONE

ALPHA

OMEGA

WELL BEING BESIDES KNOWLEDGE IS THE GOAL OF PHILOSOPHY AND SCIENCE.

GEOMETRY IS EVERYWHERE FROM THE MICRO TO THE MACRO COSMOS

ALL THINGS ARE EITHER WAVES OR CIRCLES

All Things

Conclusion:

Nature is beautiful and there are multiple things to learn by observing the multiplicity of all Natural Events. In the Universe there are good and bad aspects of all things in a Sea of Quantum Fluctuations with Waves and Circles holding in place the Matter and Energy. The Reality we currently live is formed from a collection of Waves in Hilbert Space forming the Space Time Dimension in which we live. We learn much more from observing the world outside of us than through just Reason. Experiments must be conducted to verify a Hypothesis, and to generate new Questions and make clear Statements based on Collected Data. Overthinking is a common issue in Science but the Absolute Truth is that all things can be explained with Equations and through Numbers even if restricted within Probabilities. Humans have the Potential to establish a Just Civilization ruled by Philosophers and that is free of all evil, and filled with a healthy and safe Environment in perfect Equilibrium without hurting or damaging anything in the Natural World. The Goal of Philosophy is to not only instruct us about the Natural World but also to help us reach Utopia, and a feeling of well-being and accomplishment free of suffering and with much joy.

Problems in Mechanics and Relativity

Interstellar Travel:

Suppose a Space Ship is not moving in Space. It then accelerates to 30 m/s with the Engines. The Engines are turned off and it continues moving at a constant 30m/s. For the people inside the Space Ship, it is at rest since there is no Acceleration. There is no Friction in Space so it continues at 30 m/s. The Ship then engages in Pulses where it adds 30m/s to its Velocity in each Pulse. It will take 10,000,000 Pulses for the Space Ship to reach the Speed of Light. For the reference frame of the Space Ship, it is at rest at the end of each Pulse or between Pulses, since it is at a Constant Velocity, and it Accelerates only during each Pulse. **The Space Ship's reference frame of being perceived at rest is what allows it to keep increasing its Velocity by adding 30 m/s during each Pulse**. An Inverse Pulse slowing the Space Ship by subtracting 30 m/s during each Pulse is performed to slow it down all the way to rest when near arrival of its destination.

Pulses Trick as a Means of Interstellar Travel

Observers outside of Space Ship will never see it exceeding the Speed of Light but will see it reaching ever closer to that Ultimate Speed. Observers inside the Space Ship will see after each Pulse, Space ahead of them to be closer, since the faster or closest to the Speed of Light an object moves, the more Space ahead shrinks and far away distances become small. That allows the trip through great distances not only Interstellar but also Intergalactic. Between Pulses, the Space Ship moves at a Constant Velocity and the occupants of the Ship think that they are at rest. This allows the addition of 30 m/s during each Pulse. **The biased Space Ship's Frame of Reference that thinks it is at rest, is what allows the Acceleration only during each Pulse by adding 30 m/s to its Velocity each time**. There is also Time travel to the Future since the closer one moves at the Speed of Light the slower time flows as compared to an object that is stationary. These Pulses also make the Space Ship a Time Machine. Objects at rest will age faster while objects inside the Space Ship will age slower.

Pulses Trick as a Means of Time Travel

Gravity:

When objects move near the Speed of Light they become more massive from the perspective of an Observer that is stationary.

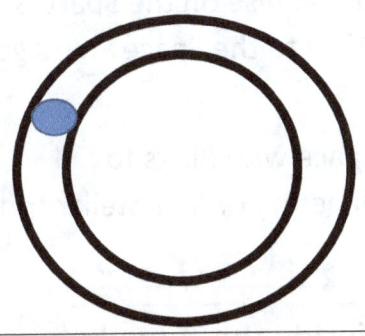

The ball can gain an infinite amount of energy, as it is accelerated by the Magnetic Rings in a Vacuum. That is because there is no Friction to slow it down and the Magnetic Field is held constant. This truly becomes a Perpetual Motion Machine.

If a metal ball is placed in a Frictionless Circular Tube with a Vacuum inside, and if the ball is accelerated by Magnetic Rings in this Vacuum, when the ball passes each ring it gains Speed similar to the Pulses explained previously for the Space Ship. If the Ball's Speed is increased significantly close to the Speed of Light, its mass will grow significantly only for an Observer outside of the Spin. This will generate Gravity pulling matter towards the center of the Circular Tube Area. If several of these Circular Tubes are placed on the floor it will generate Artificial Gravity. To stop Artificial Gravity, the Magnetic Fields of the Ring are reversed slowing the ball instead by bringing it to a stop.

Tunnels:

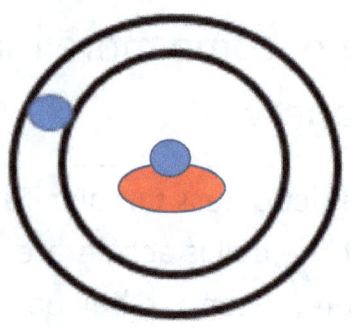

The picture above shows a Spaceship passing through the center of the ring, being pulled forward due to Gravity generated by the rotating Metal inside the Circular Tube.

These circular tubes with a metal ball rotating inside can generate Gravity at the center of the ring. If these circular tubes are placed around a Vacuum Tunnel, by having these tubes to be the Gravitational Rings around the Tunnel, it can speed a Space Ship accelerating it from the Gravity to high speeds. Each circular tube leads to a Pulse on the Space Ship passing through their center, leading to the Space Ship's gain in speed after each Pulse.

These Tunnels could be built in Space with Rings to Accelerate Space Ships to Interplanetary or Interstellar trips.

To slow down the Space Ship inside the Tunnel, the Space Ship releases energy in the opposite direction as its motion with Pulses bringing it to a stop after several Pulses. Anti-Gravity may not exist, so in order to slow down the Space Ship, it needs to exert a Force in the opposite direction of its motion. The Circular Rings, however, require no fuel from the Space Ship to speed them up. Fuel is only used to slow down the Ship.

Circular Tube with Magnetic Rings:

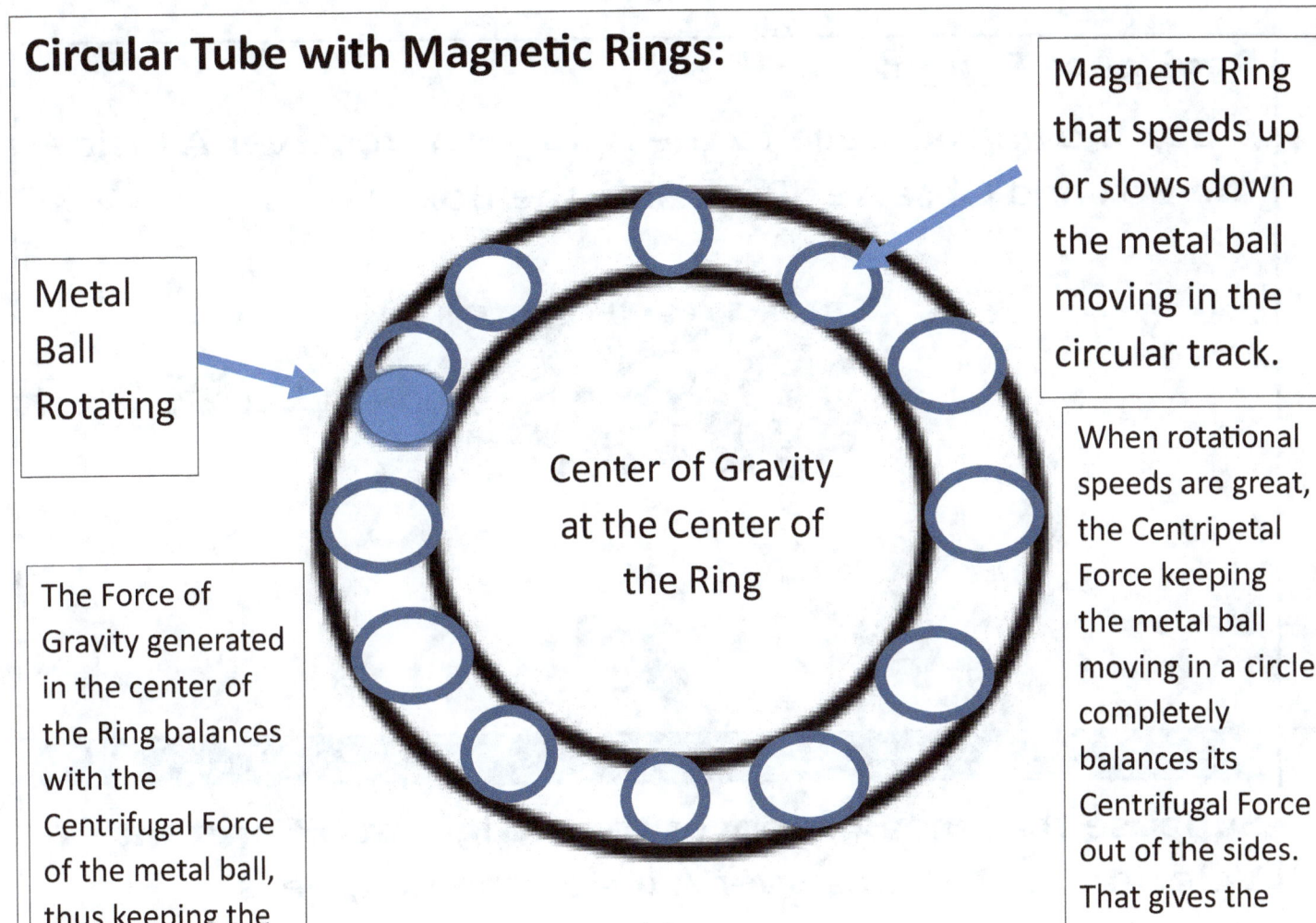

Metal Ball Rotating

Magnetic Ring that speeds up or slows down the metal ball moving in the circular track.

Center of Gravity at the Center of the Ring

The Force of Gravity generated in the center of the Ring balances with the Centrifugal Force of the metal ball, thus keeping the system stable.

When rotational speeds are great, the Centripetal Force keeping the metal ball moving in a circle completely balances its Centrifugal Force out of the sides. That gives the system stability.

Reference Frame:

A Box moving in space to the right with Observer A inside the Box and Observer B outside the Box.

Suppose that the Box moves to the right at a Constant Velocity. For the Observer A inside the Box, he is stationary. For Observer B, Observer A is moving to the right.

Let us say that Velocity that the Box moves to the right is 10.0 m/s.

In the Observer's B Reference Frame, the movement of the Box and Observer A is:

X = Vt

Where X is the Displacement, V the Velocity, and t is the Time.

In Observer's A Reference Frame, he does not move according to the following equation:

$X' = X - Vt$

Suppose that the position of Observer A inside the Box is 5 m, and the Box is 10 m long:

5 = 5 + X − Vt

Where X = Vt.

Notice that being X = Vt, and Vt -Vt = 0, the Observer A is then not moving inside the Box, he does not move in the Box Frame's of Reference.

For Observer B at 1 second the Box moves 10m, after 2 seconds it has moved 20 m, at 3 seconds it has moves 30 m. For Observer A, he is stationary through the following:

$5 = 5 + 10(1) - 10(1)$

$5 = 5 + 10(2) - 10(2)$

$5 = 5 + 10(3) - 10(3)$

Below are the Galilean Transformations for Reference Frames:

Coordinates:

Where X' is the Reference Frame A and X is Reference Frame B. All letters with the ' at their top right refer to the Reference Frame of the moving object.

$X' = X - Vt$

$Y' = Y$

$Z' = Z$

$t' = t$

Einstein generalized these relative concepts by stating that there is no universal frame of reference. Each observer will measure the universe from its own frame, and there is no absolute description of reality. Everything is relative while clocks and other measurement devices will not always agree on the same results.

Relative Velocities:

The equation for Relative Velocity is:

$$V = \frac{V1+V2}{1+\frac{V1XV2}{c^2}}$$ Relative Velocity can never exceed the Speed of Light.

Example 1:

Object A moving at 0.6c towards object B, that is moving towards A at 0.9 C.

Notice that 0.6 + 0.9 = 1.5 which means that they would be expected to be moving faster than the speed of light towards each other, but that does not happen. Their relative velocities do the following instead:

$$V = \frac{0.6c+0.9c}{1+\frac{0.6cX0.9c}{c^2}}$$

Their Relative Velocity is: 0.97c which is just below the Speed of Light.

Example 2:

Object A moves at 0.6C in one direction, and Object B moves at 0.9C in the same direction. Their Relative Velocities would be expected to be 0.9c – 0.6c but shown below is what happens instead:

$$V = \frac{0.6c-0.9c}{1-\frac{0.6cX0.9c}{c^2}}$$

Their Relative Velocity is -0.65c which is above the -0.3c that would be expected in Classical Mechanics.

Example 1:

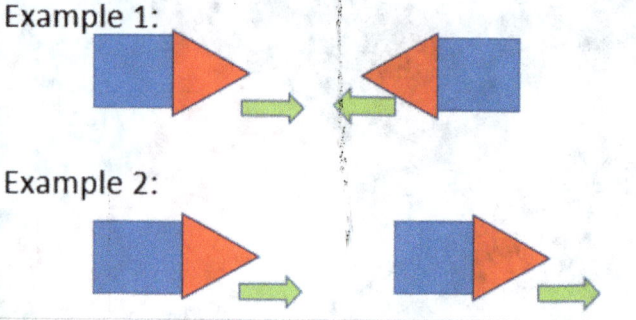

Example 2:

Muon Decay Experiment:

Muons decay with a Half Life of $2.2\mu s$. If they move at 0.9c then their new Half Life Time becomes:

$$\gamma = \frac{1}{\sqrt{1 - 0.9^2}} = 2.29$$

2.29 X 2.2 = 5.0 μs instead.

Time and distance get shorter for objects moving compared with objects stationary. That explains the longer half Life Time of Muons since their clocks are moving slower. The Muon will see it covering a distance, while for objects stationary, the Muon will have covered a longer distance and will have lasted for a much longer time.

Example:

1x10^9 Muons are travelling a distance of

10,000 m.

The Muons are travelling at 0.9c.

It will take these Muons from the laboratory frame of reference, 3.70x10^-5 s seconds to cover the distance in about 7.4 Half Life Times. Only $5.92x10^6$ Muons would be left when arriving at the distance.

$(10,000)/(0.9(3x10^8)) = 3.70x10^{-5}$ s

$(3.70x10^{-5})/(2.2x10^{-6}) = 7.4$ Lifetimes

$(1x10^9)(0.5)^{(7.4)} = 5.92x10^6$ Muons left.

That is what is seen in the experiment.

Their Half Life Time is measured to be:

$\gamma = 2.29$

And $2.29(2.2) = $ 5.0 μs

Which means that there will be only 7.4 Half Life Times and about $6x10^6$ Muons will be left when arriving at the distance instead of 17 lifetimes that was expected.

$(3.70x10^{-5})/(2.2x10^{-6}) = 17$ Lifetimes

For the Muon their half Life is still 2.2 μs, but the distance they see covering is 10,000/2.29 = 4367m. The time that the Muon measures for the trip is (3.70x10^-5 s)/2.29 = 1.62x10^-5s

Which is the same as 4367/(0.9(3x10^8).

Time and length of space are observed to be smaller for the moving Muon. It is the fact that Muons are seen to live longer and move larger distances before their decay that provides evidence for the Special Theory of Relativity. Because the Muons see themselves moving at a shorter distance and in less time, they agree with the Laboratory Frame of Reference on the number of lifetimes as shown in: (1.62x10^-5)/(2.2x10^-6) = 7.4 Lifetimes

Below is the Space Time for one of these Muons compared with the
Space Time for the observer outside the Muon.

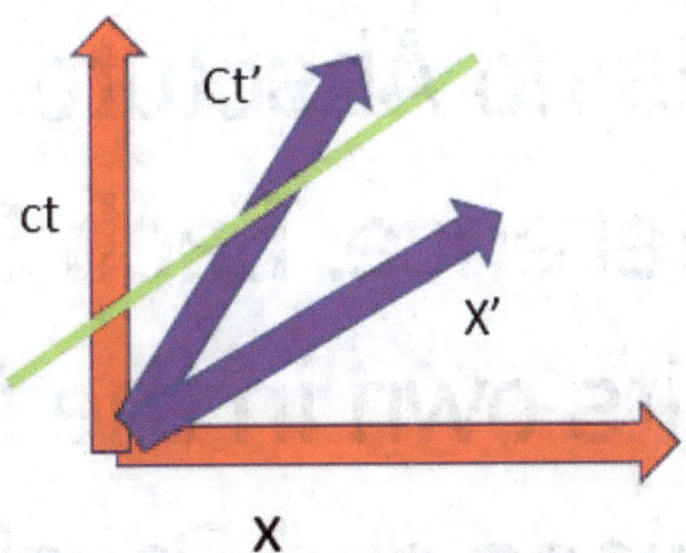

ct and x stand for the time and space of the observer of the Muon, while ct'
and x' stand for the time and space of the Muon. The green line indicates that
the observer of the Muon sees one event in time at the intersection of the
green line with ct. The Muon, however, has the green line intersecting at a

longer length at ct' which means that it sees the event happen later. This means that the observer sees the Muon's Clock tick slower thus taking a longer time to see that specific point in time. The green line has to be parallel to the x' so that this comparison can be made.

There is no Absolute Frame of Reference. Each Space Time is its own in the Multiple Dimensions and Possibilities in Space.

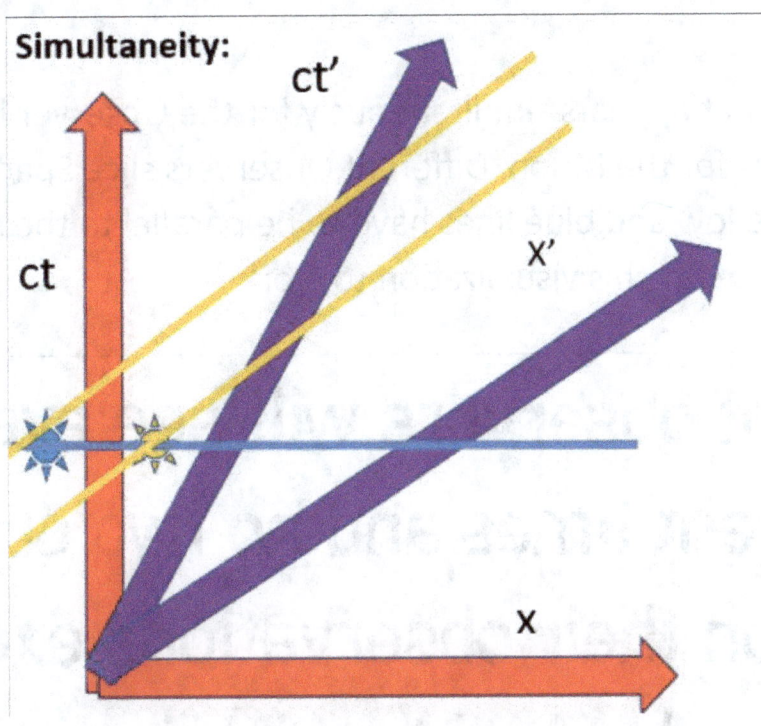

In the last image the Observer at rest sees two events shown as Blue and Yellow Stars occur at the same time in its ct time frame. The Muon, however sees the event of Yellow Star before the event Blue Star in its own Reference Frame. This means that what appears simultaneously for the Observer happened in different times for the Muon. Different Observers slice Space Time at different angles. The yellow and blue lines have to be parallel to the space line of a given reference frame for this visualization to work.

Different observers will see events happen at different times and no two observers will agree on their observations except for the Speed of Light which is a constant regardless of the Space Time reality that an observer is in.

Using the Equation:

$$E^2 = p^2c^2 + m^2c^4$$

Example:

The Rest Mass of the Electron is:

$0.511 \text{Mev}/c^2$

If its Momentum is:

1.0 MeV/c

Its Total Energy is:

$$E^2 = 1^2 + 0.511^2$$

E = 1.12Mev

Photons have no mass so their Energy is:

$$E^2 = p^2c^2 \qquad E/c = p$$

Which means that despite having no Mass, Photons have Momentum.

$P = h/\lambda \qquad \lambda = h/p$

Using de Broglie's Equation

Kinetic Energy of a Particle is:

Ek = $m_0(1 - \gamma)c^2$

Momentum and Mass:

$$p = \gamma m_0 v \qquad m = \gamma m_0$$

Energy is: $E = \gamma m_0 c^2$

The faster a Particle moves the more Mass it has.

The Twin Paradox:

Twin A enters a Space Ship and travels at 0.999c Towards Alpha Centaury which is 4.4 Light years away. Twin B stays on Earth. For Twin A, Alpha Centaury is only 4.4/22.4 = 0.196 light years away, or in other words: only 71.7 light days away. It takes Twin A only 71.7 days to arrive at Alpha Centaury and another 71.7 days to return to Earth at the same Velocity of 0.999c. When both Twins meet, Twin A has aged only 71.7 x 2 = 143.4 days, and Twin B has aged 4.4 x 2 = 8.8 years.

The rotating metal ball inside the Circular Tube Ring can generate such an intense Gravitational Field so as to become the entrance to a Wormhole allowing a trip back or forward in time.

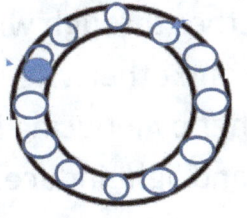

An example:

Entrance in the year 3000

Warped Space Time due to intense Gravitational Field from the rotating metal ball connecting two points in time.

Wormhole

The equation for the strength of the Gravitational Field at the center of ring is **proportional** to:

$$g = \frac{\gamma v^2}{r}$$

Ring

Exit in the year 1986

Where g is the Acceleration due to Gravity, γ is Lorentz Factor, v is the speed of the metal ball, and r the radius of the ring.

In the same way that by inserting or removing a Magnet inside a coil, a current is generated in the coil, the motion of up and down charges along the Antenna can only lead to light propagation in Space, if Space is made of something rather than nothing. In the same way that a Magnet can only generate a current if there is a coil, the light coming from an Antenna can only propagate if Space is made of an Ether through which light can flow. This Ether is only detected when light moves through, and otherwise it behaves empty or made of nothing. Light propagates through the Ether by inducing its flow through the Ether Space. A Magnet induces a current in a coil, and light induces its flow or propagation through Space.

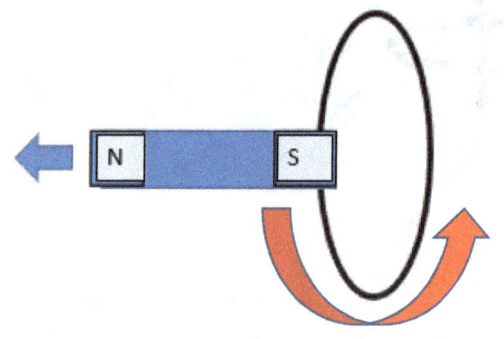

Oscillation of Charges in an Antenna induces the flow of Light in the Ether of Space. The Ether does in fact exist.

Motion of a Magnet induces a current shown in red on the coil.

What if?

What if there is in fact such thing as Anti-Gravity? What if Anti-Gravity can be generated by simply reversing the direction that the metal ball rotates inside the circular tube or ring. Example shown below:

On top of the ring, the Gravity
pulls matter in the wormhole

Is the entire Universe inside a Gravitational Dipole?

Wormhole

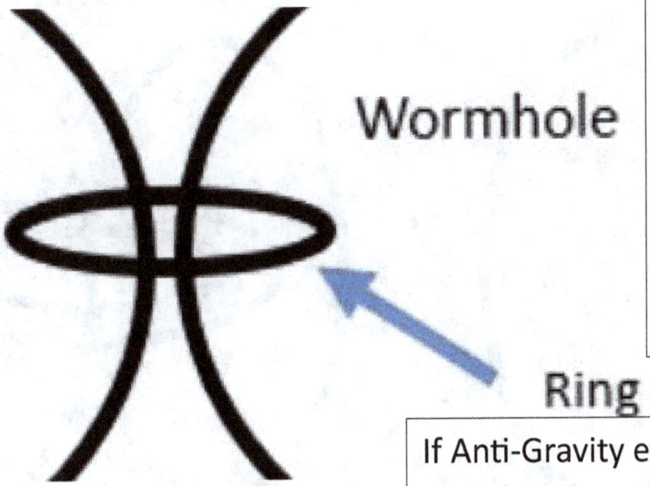

Could a Wormhole be a Gravitational Dipole just like a Magnet is a Dipole and two opposite charges generate an Electric Dipole?

Ring

At the bottom of the ring, Anti-Gravity spits out matter

If Anti-Gravity exists, then the rings can speed up Space Ships or slow down Space Ships travelling inside Wormholes.

178

Similar to a Magnetic Field and an Electric Field?

In the same way that a current going through a coil generate a Magnetic Field, could a rotating mass inside a ring at high speeds generate a Gravitational Field that is attractive on one end and repulsive on the other?

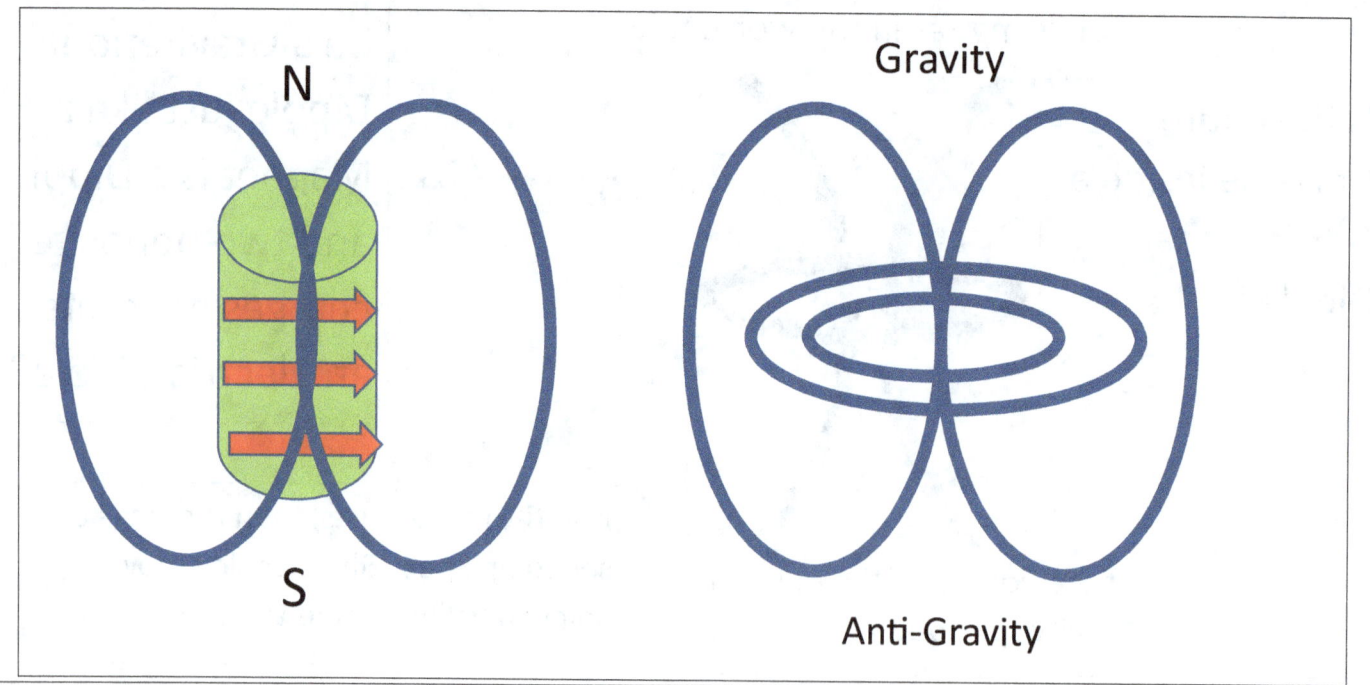

The Equation:

The Gravitational Field Generated in the Center of Ring with the metal ball rotating is proportional to:

$$g = \frac{\gamma v^2}{r}$$

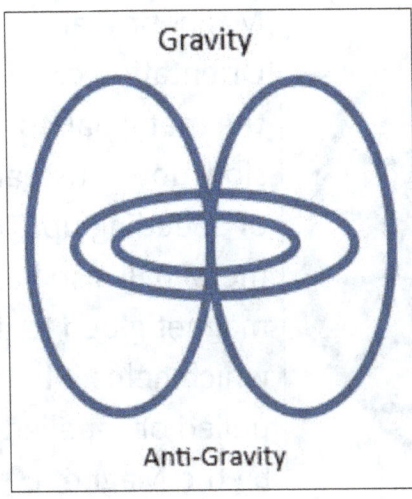

Gravity

Anti-Gravity

The more massive the metal ball is, the greater that the Gravitational Strength will be. That means that the complete equation is:

$$g = \frac{\gamma v^2}{r} (MX)$$

Where M is the rest mass of the metal ball and X is a Constant of Proportionality that when placed in the Equation it gives the Gravitational Field Strength g.

g is the Acceleration due to Gravity in the center of the ring.

The metal ball rotating inside the ring can never exceed the Speed of Light, but the closer it gets to the Speed of Light, the more massive it becomes in generating Gravity in the Center of the Ring. The more massive and the faster that the ball moves, the more it can warp Space Time, and the more it can pull or push matter in Space.

The metal ball is accelerated inside the Ring by Magnetic Rings around a Vacuum Tube. Because the tube is a Vacuum, no Friction exists to slow the ball.

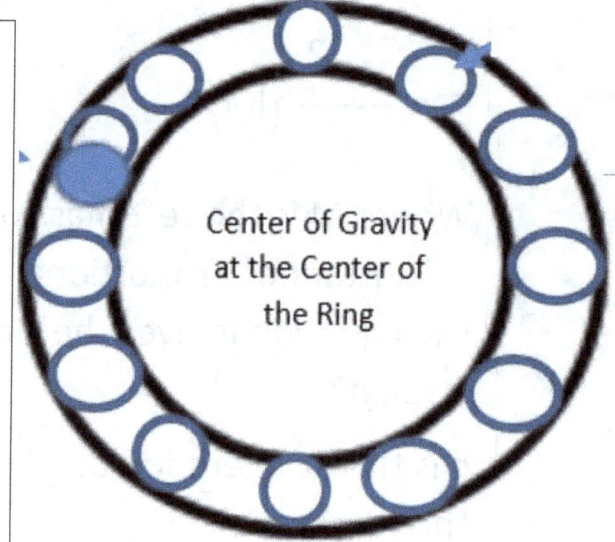

Center of Gravity at the Center of the Ring

The Magnetic Rings can reverse the Magnetic Field Orientation causing the metal ball to slow down instead of Speeding up. The metal ball also has a magnet glued to it which helps it being pulled or repelled by the Magnetic Rings.

Unifying Gravity with the other Four Forces of Nature:

If an Electric Field pushes charges in a coil, generating a Magnetic Field from rings around a Circular Tube, that pushes a metal ball through those rings accelerating it to a high speed, and the Circular Motion of the metal ball generates a Gravitational Field in the center of the Circular Tube, then it is proven the connection that the Electric and Magnetic Fields have with Gravity.   This thought experiment proposes the conversion of Electricity into Gravity.

All Four Forces of Nature are Connected

Anatomy of the Circular Tube

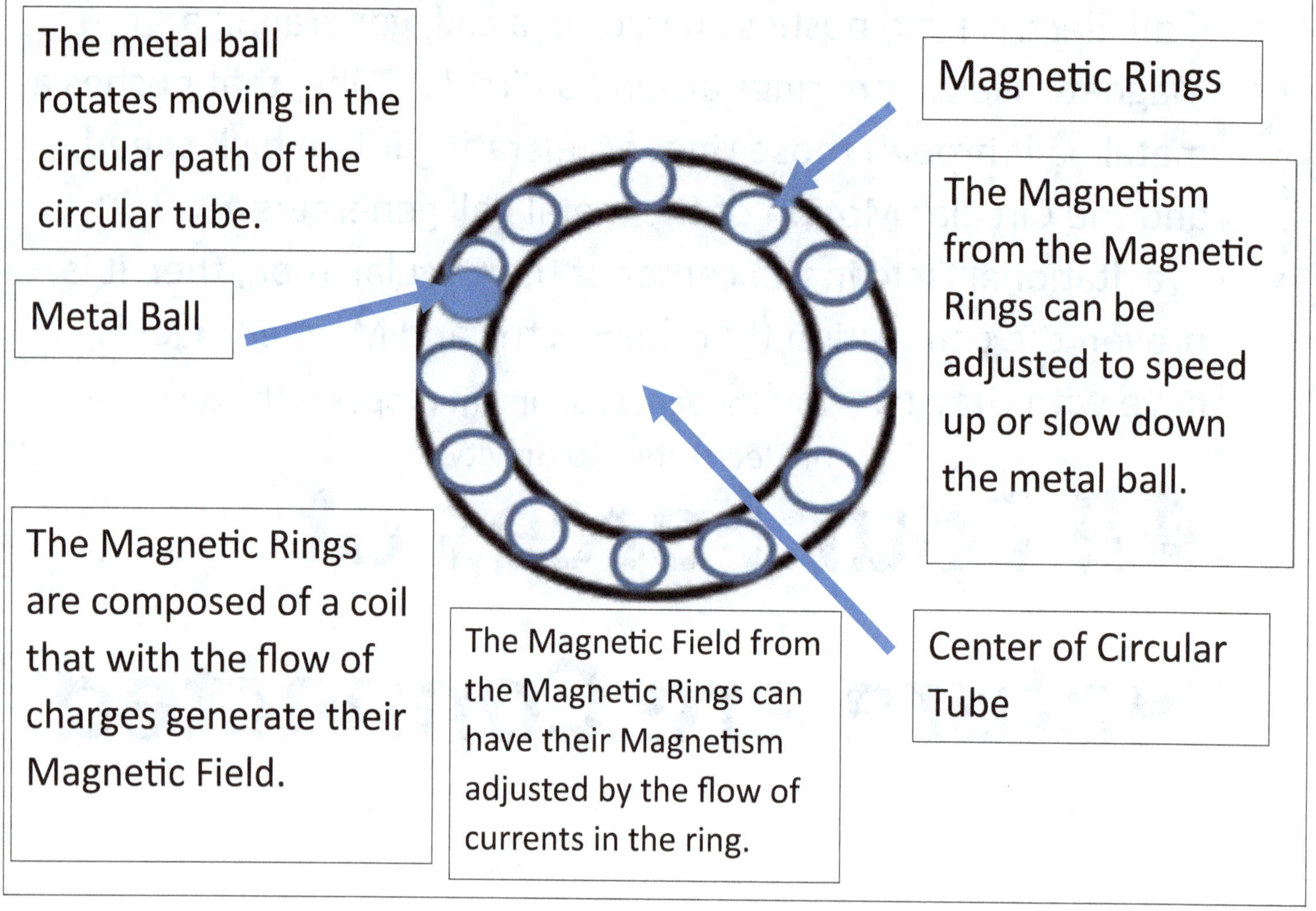

The metal ball rotates moving in the circular path of the circular tube.

Metal Ball

Magnetic Rings

The Magnetism from the Magnetic Rings can be adjusted to speed up or slow down the metal ball.

The Magnetic Rings are composed of a coil that with the flow of charges generate their Magnetic Field.

The Magnetic Field from the Magnetic Rings can have their Magnetism adjusted by the flow of currents in the ring.

Center of Circular Tube

A Wormhole Tunnel could be built with several of these Circular Tubes like rings around the tunnel that pull or pushes matter in or out of the Tunnel.

Each of these rings are a Circular Tube as explained in the previous page.

These Circular Tubes are Rings that speed up or slow down objects going through the Wormholes Tunnels.

A Magnetic Ring is made of a Coil in the form of a Toroid that when Current goes through it generates a Magnetic Field. The direction of the Magnetic Field can be changed with the change in the direction of the Current through the coil that can speed up or slow down an object moving inside the Circular Tube.

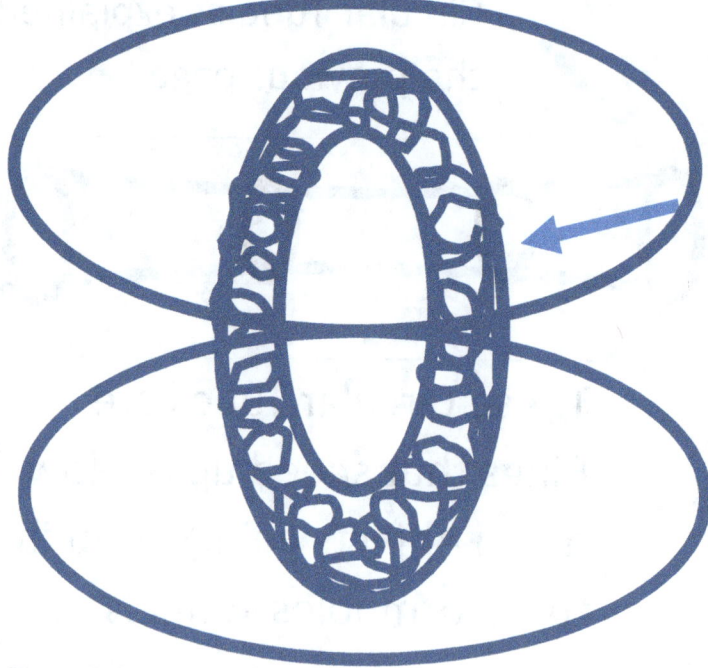

These rings speed up and slow down the metal ball inside the Circular Tube which generate Gravity.

Coil in the shape of a Toroid generating a Magnetic Field in the Center of the Ring from the flow of Current in the Coil.

Feynman Alternative Histories state that there are a great number of Possible Histories that led to our present and that will lead to our future. Because of Feynman Alternative Histories, you could literally go back in time to a Parallel Reality to prevent your parents from meeting each other, but you would still exist in a Parallel World. Some people argue that this would not be possible since in preventing your parents from coming together you would then not exist. Since the Universe has multiple Histories and Parallel Realities, you can go back in time to a Parallel Reality, change that reality, and History will take a different route with you as part of this new and modified reality among multiple in the cosmos. It is this great number of Histories and Dimensions that allows travel back in time and even the possibility to change a specific History among multiple Possibilities in the Chaos of Space Time.

Histories and Dimensions Explained:

Suppose that there are 3 Parallel Universes with three different histories:

History 1:   A meets B and both lead to C.

C then goes back time and changes History 1 forming History 2:

History 2:   A meets C and A does not meet B, C still exists.

C then goes back in time again to meet himself preventing himself to interfere with A and B.

History 3:   C meets C allowing A to meet B and both lead to C.

There are now 3 C's.

All three histories are Parallel Realities that differ from each other obeying the many Possible Ways things can happen according the Feynman Alternative Histories. This shows that going back in time is possible and that going back in time can even multiply the number of yous in a Universe.

Multiplying the number of yous. Another form of Cloning:

Say you go back in time to two days ago. You meet yourself and now there are 2 of you. Two days later both yous go back in time to meet the 2 of yous in the past and now there are 4 of yous total. If you keep doing that, you are literally cloning yourself and multiplying the number of yous. Since each Electron is identical to another Electron, and each Particle is identical to another Particle of the same kind, then maybe all the great number of Particles in the Universe were created by making copies of themselves using time travel as described above.

If you instead travel to the future, you will not see another yourself simply because you left the world in the past. If you travel to the future you will hear from people that you disappeared in the past and now is back. On other words: cloning can only happen by travelling back in time and not forward in time.

Using Time Travel to the Past in order to Clone Yourself

Cloning Particles:

Let us say that we place a single Particle inside a Wormhole that curls on itself. The Particle moves through the Wormhole and meets itself again in the past. Now there are 2 Particles. They both go through the Wormhole and they meet themselves again in the past. Now there are 4 Particles. These 4 Particles go through the Wormhole and meet themselves in the past again. Now there are 8 Particles. This thought process proves that to create the entire Universe all you need is one Particle and a Wormhole. The Particles can then combine with each other forming other Particles and the more that these Particles go through the Wormhole more copies and the greater is the Multiplication of them throughout Infinity. I believe that all Particles are made of Photons, and Light is Energy, and from Energy comes all the Matter of the Universe according to the Equation $E=mc^2$. Energy can also multiply itself several times starting with just a single Photon inside a Wormhole. Let there be Light, and creation was brough forth!

An entire Universe from
Nothing!

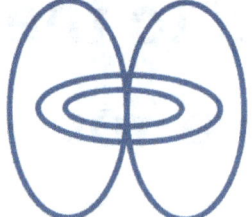

Revolution of World Progress

Civilization

Importance of Human Organization:

It is clear and evident the amount of organization seen in the natural world. The complexities at the Atomic and Molecular Level that provides the structure that sustains reality, is an object of wonder. Humans that are aware of these complexities, have also created with the advantage of these Laws and Constants, all the Technology we currently have. Although Technology and Science has proven how much within the cosmos can humans have control, it is also evident that the majority of the world population is completely ignorant of these advances in Science. In the same way that each part of the universe has a function, each part of a cell phone also has a function, which means that each human must fulfill his or her function in the society. It is necessary that everyone understand their purpose in the human world. Our current civilization shows that a new structure needs to be built, to organize humanity. The present structure of civilization is not being very useful since there is a growing chaos in the society especially in education and the human moral values are slowly being lost in time. Technologically we keep advancing but our human soul and moral values are being brought down and being lost. We are losing our identity and purpose in life.

The Human Purpose:

Our humanity in the current world is gaining a mindset that is highly destructive. Education is losing its purpose which is to educate, to challenge, to lead people to grow in knowledge. No wonder that the Ancient Greek Philosophers were highly preoccupied with politics and were not only thinking about Science. Socrates is an example of a Philosopher who was not worried about Science since he believed that without Politics and correct human morality, civilization would not be able to keep itself standing for too long. Science is tremendously important, but I also believe that without good politics, and without a well structured educational system, civilization collapses, and then Science stops advancing. What leads to advances in Science is a civilization that values the Principles required in Politics, that has a positive effect on Education, and in the developing of several generation of citizens, that bow down to a common flag, that accept the preserved moral values, and that abide to the mindset of growth and progress. It is this mindset of growth and progress that is diminishing in the current world and this is quite worrisome since it could lead to dictatorship, overthrow of current world order with a war, rebellion, and a complete disaster which we should avoid.

The Wrong Pyramid:

There is a wrong mindset present in society today, which is causing ignorance to be much more common than knowledge. The few who know are leading to advances in Science and Technology, and the majority that do not know are blindly following Principles that keep them ignorant and in lower social classes. There are clicks in pop culture that stimulates the thinking on people that they are no capable, that they are weak, and even that they are worthless. This negative thought about themselves leads them to not go to college, to not believe in themselves and to just follow directions without questioning their existence and purpose in the society. It seems that there is an invisible caste system that instructs people through the pop culture to continue ignorant. I say it is invisible because there are plenty of opportunities in education, and career, but the culture generates clicks in people that guides them in the wrong paths, leading to rebellion, and them to be not willing to stand up for their national anthem, or to praise their country's flag. If ignorance becomes even more common than knowledge, society will be sustaining this invisible caste system that will lead to fewer and fewer people in power until the whole civilization collapses.

One Person can't do it Alone:

Socrates was one man, one of the Ancient Greek Philosophers, and he became one of the most important, becoming a reference for many other people ever since throughout world history. Ancient Greece was the first of all European Civilizations, leaving a legacy for the entire world and an influence still very present in modern times. One person can not do it all alone, however. Albert Einstein discovered the Theory of Relativity, and Isaac Newton discovered Classical Mechanics, but if it was not for other people to spread their advances, these scientists would end up forgotten or unknown. It is worth thinking whether in world history was there exceptional people that never made to popularity and became forgotten. Could it be that there is a complete genius that is so smart but that never shares his or her knowledge and then dies with all that information only for her and himself? Sometimes high level of intelligence leads to struggle with social interactions leading to isolation and no other people meeting that genius and no world progress in Science is made from that person. This is why one person can not do it all alone, but only those who shares their thoughts and imagination to the rest of humanity.

The Order of the Society:

Below is a description of the Flow and Order that keeps the Civilization Harmoniously and Functioning.

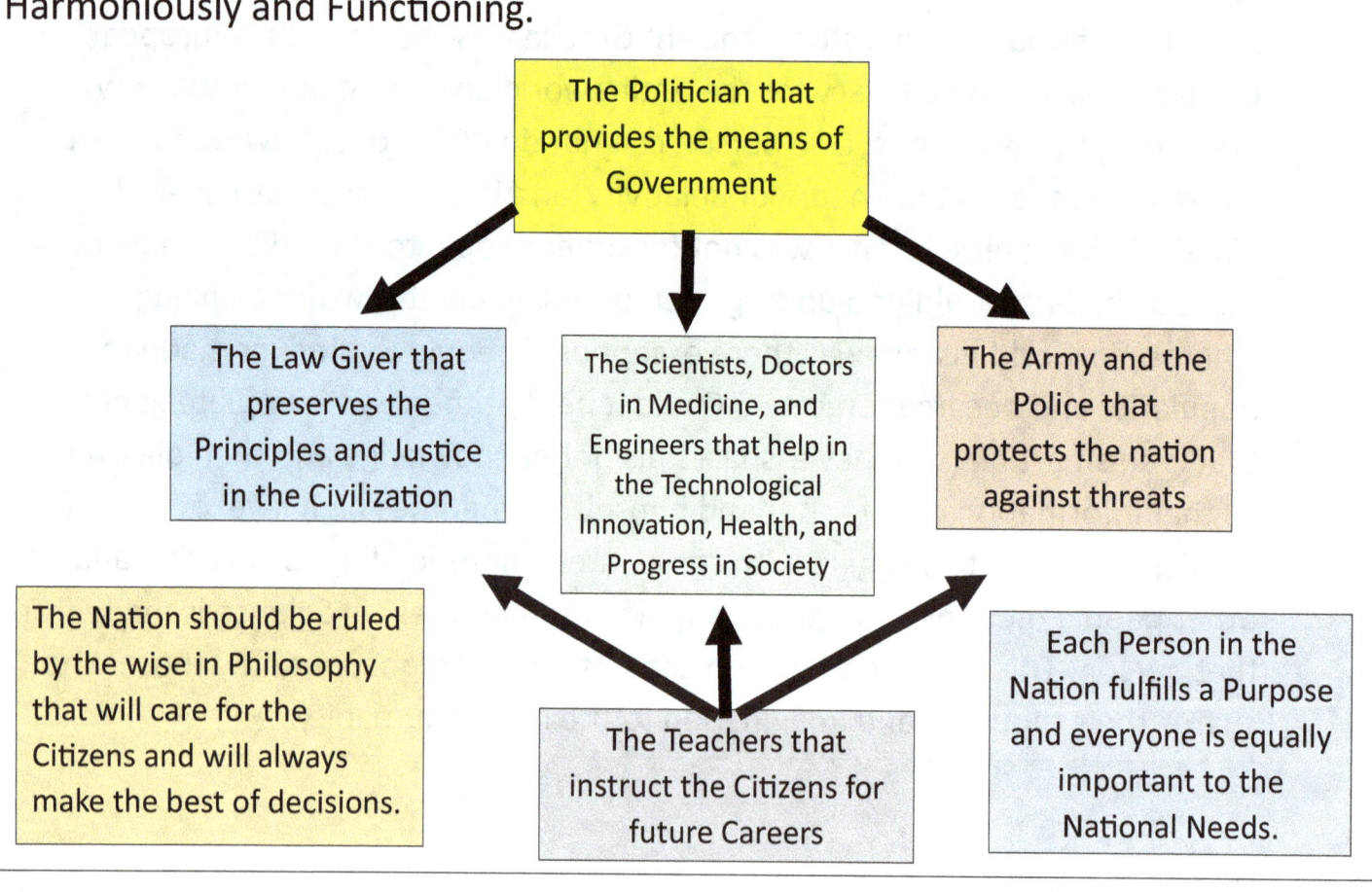

The Politician that provides the means of Government

The Law Giver that preserves the Principles and Justice in the Civilization

The Scientists, Doctors in Medicine, and Engineers that help in the Technological Innovation, Health, and Progress in Society

The Army and the Police that protects the nation against threats

The Nation should be ruled by the wise in Philosophy that will care for the Citizens and will always make the best of decisions.

The Teachers that instruct the Citizens for future Careers

Each Person in the Nation fulfills a Purpose and everyone is equally important to the National Needs.

The Order of Science:

Here is the Map of the Flow of Scientific Thought in the Society.

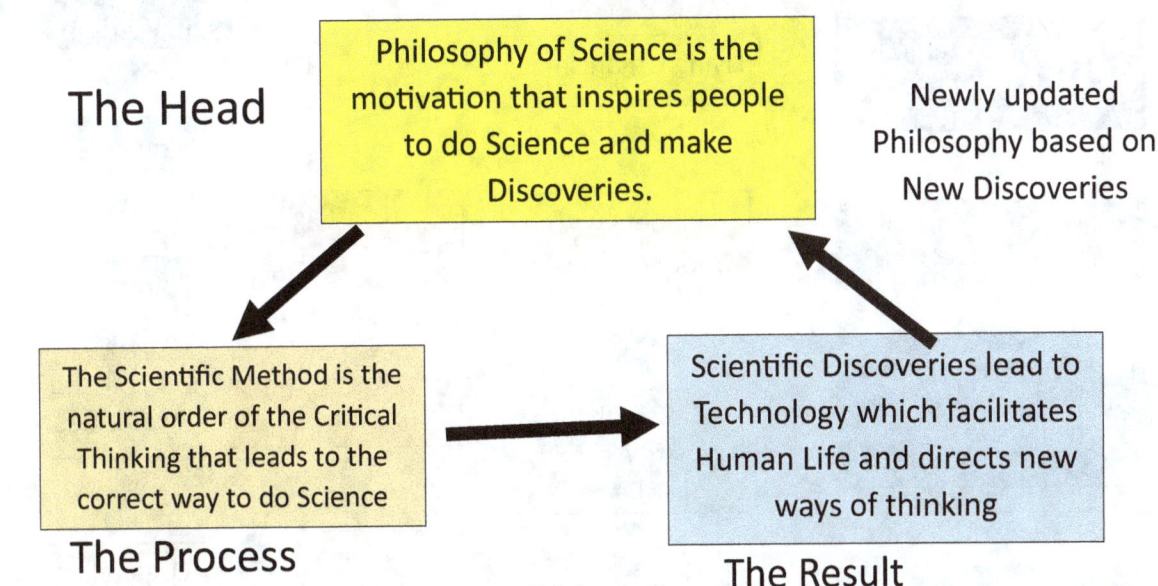

The Head

Philosophy of Science is the motivation that inspires people to do Science and make Discoveries.

Newly updated Philosophy based on New Discoveries

The Scientific Method is the natural order of the Critical Thinking that leads to the correct way to do Science

The Process

Scientific Discoveries lead to Technology which facilitates Human Life and directs new ways of thinking

The Result

It is not possible to start Science without the Philosophical Mindset and without knowing what Science means and the benefits humans acquire from the art of knowing. The Scientific Method is the established order of events that is comprised of steps needed in order to perform a true and valid approach into the exploration of reality. The Results may be new discoveries and a Newly Updated Description of the World we live in. The Results lead to New Technologies and forms of living that are only good and valuable if improves the human wellbeing and that is not destructive to the natural environment. The goal of Science is to guide the healthy progress of civilization where all people can benefit from it.

The Map of the Universe:

Nature has an order and is comprised of several levels from greatest to the smallest, each with its own function.

| Multiverse comprised of all Universes |
| → |
| Universe comprised of all things in a Cosmos |
| → |
| Cosmic Web with multiple Clusters of Galaxies |
| → |
| Galaxy Clusters of several Galaxies |

Galaxy Clusters of several Galaxies → Galaxy with all Stars, and Globular Clusters → Stars with the Planets, Asteroids, and Comets → Planets with Moons and Life → Living Organisms → Souls → Brain and Organs → Organic Tissues → Cells → Organelles → Molecules → Atoms → Particles → Quantum Fluctuations → Wave Particle Duality

The Map of a Human Being:

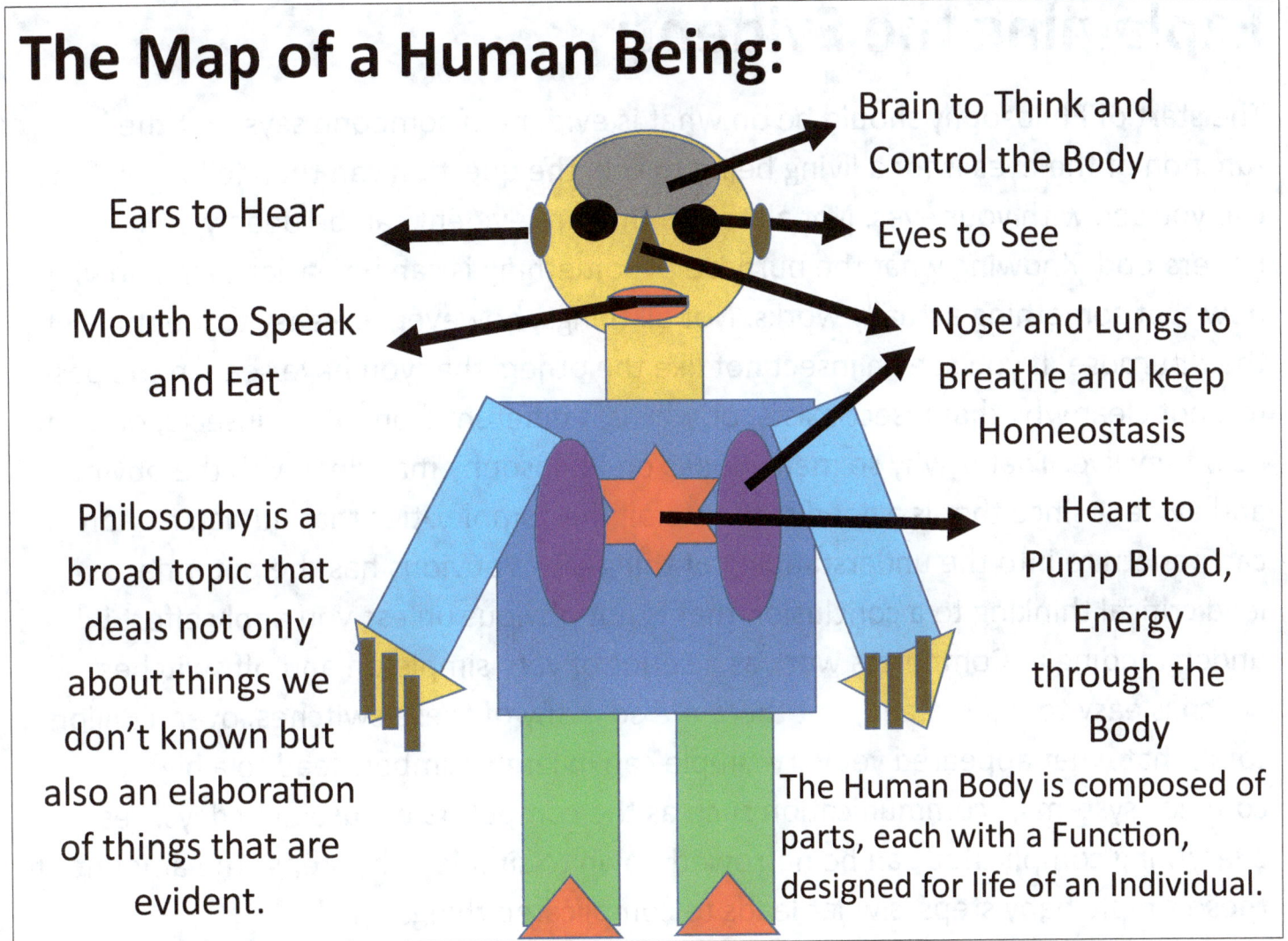

Brain to Think and Control the Body

Ears to Hear

Eyes to See

Mouth to Speak and Eat

Nose and Lungs to Breathe and keep Homeostasis

Philosophy is a broad topic that deals not only about things we don't known but also an elaboration of things that are evident.

Heart to Pump Blood, Energy through the Body

The Human Body is composed of parts, each with a Function, designed for life of an Individual.

Explaining the Evident:

The start of Philosophy should be on what is evident. If someone says that the function of the eyes is for a living being to see, the question can then follow on how can you see with your eyes. Not all things that are evident can be deeply understood. Knowing what the purpose of something is can be easier than knowing how that something actually works. Not all things, however, are easy to identify with their purpose. If you see an insect not like the others that you have seen in the past, it is not clear why that insect exists, or why is it different from other insects, or even how they live. That is why so many books on Philosophy may deal with the obvious and evident since that is a good start into all the complexities that human reason can penetrate into the understanding of things. The obvious has the potential to lead critical thinking to a conclusion that is not obvious unless you apply effort in understanding it. Computers work as a series of very simple on and off switches, which is easy to understand, but there are so many of these switches, over a Billion total, that what appeared yet very simple can in large numbers lead to a highly complex system of communication such as the computers we use now days. Yes everything complicated can be narrowed down to simple baby steps. The amount of these simple baby steps is what leads to complicated things.

Baby Steps:

All complicated things are composed of Baby Steps.

In solving the equation: $\dfrac{6\sqrt{x+9}}{2} = 3$ a person needs to know some simple rules of Algebra.

The first is to square both sides to get rid of the square root.

$\dfrac{36(x+9)}{4} = 9$ Next divide 36 by 4.

$9(x+9) = 9$ Next is to divide both sides by 9.

$(x+9) = 1$ Next is subtract both sides by 9.

$X = -8$ The answer is x = -8

All Physics and Math Problems either complicated or not can be broken down into simple known laws and can be explained with small steps. These small steps when placed together comprise the whole complexity of equations, machines, and all of our understandings of the natural world. It all starts with something evident, or a simple statement of a fact, and these simple facts over time during Critical Thinking, gains complexity, leading to the great marvel of the elaborate Theories devised by Scientists and Engineers. The Universe evolves from something as simple as a single Particle of Light, to the Immense Galactic Web Structure that holds multiple realities and Dimensions in place. It all started with a tiny Singularity and over a very long time evolution led to all we see.

Philosophy from Mathematics:

Philosophy can be derived from all areas of learning since the definition of the word Philosophy is Love for Wisdom, and Wisdom is Sophia which is Knowledge. It is possible to make Philosophy out of the simple equation demonstrated in the last page. Algebra stands for balancing two sides of the equation. Such as x = x.

If we have the following statement: 9 = 9, it is proven true since a number is equal to itself and 0 + 9 = 9 is also true since a number plus zero is itself. An equation has two sides on opposite sides of the equal sign. If you want to keep the two sides the same, whatever application you do on one side must be done in the other. So if I add 2 to 9 I must do it on both sides so that both sides can continue equal. Algebra is keeping a balance in between weights. The weights are the numbers. 2 + 9 = 9 + 2.

In solving the equation: x + 2 = 9 we see that both sides are equal to each other. What value of x will then preserve the equality? We can solve for x by leaving it by itself on one side. We can only leave x by itself if we get rid of that 2. We get rid of that 2 by subtracting 2 and that must be done on both sides to keep the balance.

X + 2 -2 = 9 -2

X = 7 the answers is x = 7

By keeping the balance between two equations, it is possible to solve for the missing variables. In Algebra there will always be something equal to something. It is the fact that they are both equal that leads to a solution, or several solutions for a given problem. A problem can have no solution, one solution, two solutions, and so forth all the way to an infinite number of solutions.

The equation: $x^2 = 36$ has two solutions which are x = 6 or -6.

Since two negative numbers multiplied to each other equals a positive number.

6 x 6 = -6 x -6 = 36

When finding roots of an equation it is possible to have one, two, three, and many solutions as many as infinity.

The equation below has two solutions.

$x^2 + x - 14 = 0$

Simplifying the equation we get:

(x-3)(x+4) = 0

With solutions x = 3 and x = -4 that when plugged in the equation equals to zero.

The following equation has 3 solutions:

$$x^3 - 3x^2 - 54x = 0$$

Which can be simplified into:

x(x-9)(x+6) = 0

With solutions x = 0, x=9, and x-6

As can be seen, the Philosophy that is possible from Mathematics is unlimited from the fact that all things are Mathematical in nature. When you say that you are happy, you may say that you are very happy, just happy, or somewhat happy. In our common language or in our thinking process and within our perspectives of all things around us, we give degrees of measurement. When cooking you are doing Chemistry, but Chemistry deals with mixtures and there is the right amount of this plus the right amount of that. Mathematics is the most fundamental of all Sciences since it is everywhere.

When playing soccer you control how fast to perform a move, where in the field to be in order to receive a pass or how to dribble when a player comes against you. All these are measurements and anything that involves measurements is numerical in nature. Even the person that does not consider him or herself a Mathematician is doing Math while driving, when choosing a clothe depending on the measurement of how cold it is, or which combination of colors to wear. Colors are just different frequencies of light which is also Mathematics. Is there the number for beauty and fine proportions? All things in the cosmos are just different regions of the same Spectrum. Just like all forms of light are just different frequencies but are still light, maybe all humans are the same exact being but each soul is a specific Frequency just like each Radio Station is a specific Frequency. The Spectrum of the whole universe is where everyone is located. Is it possible to map the soul? In mapping the universe we are mapping the entire Spectrum and an infinite number of possibilities. Mathematics is in our soul, in our perceptions, in everything we do. Mathematics is the Language of the universe. Everything comes in discrete amounts. Now being all this within measurement, and all things understandable through reason, how can we create an international order that is the most beneficial for the civilization?

Ruling Humanity:

In the same way that Algebra is a Mathematical word that revolves around the concept of balance on both sides of the equal sign, life in general is also about keeping a balance. Waves are all oscillations, of up and down motion. If it goes up it will go down, and if it goes down it has to go up, like a rubber band that stretches or a spring that when compressed and stretched has a natural tendency to attempt to reach equilibrium again and always moving back and forth. When people spend a long time listening to music, they will later decide on doing something else. The choices we make in life are oscillations within our constant desire to reach equilibrium. When a team wins many games people desire victory from another team. Does that mean that games are worthless? World History and the world of mankind has a constant desire to reach equilibrium. Is that really how it should be? Could we all benefit from a balanced world? There are fears that excessive thoughts on the implication of a balanced life may lead to a Communist type of government. It brings to question how should we live, and how should we be ruled. If a team is tired of winning, will they lose on purpose? Is obsessive desire for equally a problem? Should we instead not care about that at all and just live and be happy? Here then begins my thought of Politics.

The Rise of a New Democracy:

If everyone is equal in heaven and no soul can be better than another, does that mean that people don't play soccer in heaven? Is competition something entirely good or just a means of survival in our world? How would the world be without competition? If we want a perfect utopian world, will humanity of the future accept social differences? What if humanity is undergoing a transition from the old world towards a new world? What if our current obsession with equality is just the beginning of something entirely new? We are an Intelligence Race of beings and maybe our current transition of thoughts and events is not something completely different from what is expected at modern times. Possibly when an Intelligent Race in the universe reaches our level of technology and Scientific Understanding, the next step into that progress is obsession for equality. Before the Industrial Revolution, humanity engaged in wars, rebellion, the fall of empires, tyranny, and competition since the world had the ideology of the survival of the fittest. As humanity made advances in Science, no longer we think this way. We don't accept one person to be inferior than the rest. Humans are now changing the world order. Is this the new democracy?

The Awakening:

Imagine a world in our future. Every person wearing a white uniform with a helmet where they were connected with every other person in the world through the Internet. It was prohibited for people to wear different clothes, and to appear different, or to compete against other. Everyone had the same white uniform and had the same helmet over their heads. This helmet was controlled by their mind, and it was their cell phone, Internet Browser, GPS, personal mentor or instructor, and all these people received orders from a Supreme Intelligence that was a living Artificial Intelligence with a large database similar to Google today. Competitions were prohibited and the obsession for equality made playing sports against the law. No longer people played sports against each other, or any game including video games that were all prohibited. All music, books, and forms of entertainment was only allowed if it had the potential to lead people to perform noble acts and that were perceived as healthy for the civilization. Everyone also earned the same amount of money, which was not the same money that humans have today. Money was called points, and a person could only use a given amount of points monthly. Money was not a paper, but rather an Electronic Card that people carried in their pockets.

The World Order:

All education and healthcare in a very advanced civilization is all free. Money which has the purpose to keep the order of the society fails to fulfill its purpose if it brings chaos to the society instead. In the example of a future civilization, I see no more money present and instead I visualize something like points. Each human being earns monthly points in their personal account that can be accessed through the Internet. This individual can purchase objects or food with those points, and the reason why these points are a fixed amount is to guarantee that an individual will not spend more than he or she should. The measure of these points may vary from person to person. Some people may need more to survive especially if they suffer from an illness or are raising children. The essential needs, however, which are education and healthcare would need to be free. No access to education leads to an ignorant population with no job opportunities, and no healthcare means the person has no way to receive treatment from a disease. **If having no money can kill you or keep you ignorant, society must re-think the purpose of money and create something else to replace it**. I picture a different world order in the future to be more compatible to the basic needs of the human population.

The Helmets:

Is there money in heaven? If the answer is no, and believing on the fact that life on earth should mirror life in heaven in order for the human civilization to prosper, then why does money exist in the world? Hunger, poverty, and wars are linked to this great evil called money. **If money was good, heaven would have money.** It may be a natural process of human progress to eliminate things that are contrary to evolution. What is going to happen in the future is not known. Will humans keep advancing in Science, or will we return to the Middle Ages, or even farther back to the ancient times. With the pace that the Human Civilization is going as of 2023, it is not unlikely that technology will become increasingly important to the point where humans will forever be connected to some form of device such as a Virtual Reality Helmet, Cell Phone, Computer, and other Electronic Devices. Is that an inevitable future? Progress requires right choices too. Whatever is beneficial and worth is what humanity should choose in its evolutionary path. That means that the other alternative is the return to the natural world and only use the Electronic Devices for Educational purposes. It is clear the evident lack of certainty about human future. Is technology entirely good and how much should we use it?

The Two Futures:

Modernity is filled with dangers since humanity has embarked on a Technological trip and there is no way back. Can we imagine a world without cars, airplanes, computers, and any electronic device? Can the world ever go back to all the green forests, vegetation, grass fields, and clean water like it was in the past? The future is very scary with the fact that in order to save the life on Earth we should attempt what is currently impossible. Even the food that we eat, has genetic modifications due to the human presence in this world. There appears to be artificial things everywhere we look. If there is not a city there is a farm or a ranch, but farms and ranches are also not natural. Even human agriculture is destructive to the environment since it may require the cutting down of trees, cleaning up of forests, and even the horrible act of enslaving animals that are domesticated to serve us for food and other resources. Even before humans existed, the world was already a very bizarre place with massive extinctions and animals eating other animals. Is our current technology another mass extinction? Is human conquest of the world with a destiny to completely transform the planet and bring total destruction? Is that inevitable? Can we even live without technology though?

What if humans are not from this world, and that is why we are helping destroy it? Below I explain more:

Humans Apart from Nature:

Mankind has adapted to the life indoors, inside dwellings, buildings with air conditioning, and a clean house. Many people can't sleep over grassland, or pass the night inside tents without being attacked by bugs, and having severe allergies. The ancient people were more adapted to the life outdoors than us. Over time, humanity is becoming more dependent on modern homes and cities for their survival. If a kid is released in a forest he or she would not know what to do. A kid in the ancient times would have a better chance to survive by knowing how to obtain water and food in a forest. It is obvious that this human dependency on artificial things to live is evidence that we no longer belong to the world. If we belonged to the world we would be living in it with deep contact with nature. We are, however, rapidly saying farewell to the old times when we and nature were one. Our ancestors had no problem sleeping between trees, and they were good hunters, and very smart in the jungle. Today we are creating bonds to computers, to the Internet, and these technologies are becoming part of our everyday life. We are no longer of this world.

Our Real Home:

In a future humans may establish colonies and cities on the Moon and Mars. People may argue that life inside bubbles of air, and complete isolation from the outside world could lead humans to madness. In Lunar and Martian Colonies, humanity will have to live inside domes and bubbles or air since there is no air for breathing on either places. What can be said is that even on Earth we no longer breathe the air outside but the air indoors which goes through air conditioning. Humans are already losing contact with the natural world outside of their buildings and homes. Life on Mars and on the Moon will be even more isolated, but just like everything in life, we can certainly adapt to it. If a person is born in a colony in space, inside a city on a Satellite that orbits the Earth, and if that same person never gets a chance to meet the Earth or any other planet, that individual will just simply be completely adapted to that isolation and will not even complaint about it. Living organisms can easily adapt to changes. The Industrial Revolution happened in no more than 200 years ago and we are unable to imagine life without these technologies. A cell Phone is less than 30 years old and we already forgot how life in the world was before it.

Humans adapt to changes fairly quickly and our new home no longer is the forest but indoors, in domes, in buildings, in isolation from the outside world.

Lack of Creativity:

One observation that can be made in today's world is the populations' lack of creativity. It would be expected that so much access to technology would lead to a widely diverse content and explanation about everything. In fact, that does exist as can be seen in Internet Search Engines such as Google and Yahoo. Several people have videos on YouTube, and access to information is much easier and we do it every time we use our phone and computer. Despite all of that access, humanity suffers from lack of creativity that is far worse than in the past before the Internet. The conversations between adults never goes past the same topics, complaint about the very same politics, or how the world was better in the past and is horrible now. There seems to be a lack of topics to talk about, and there is a severe limitation on what we do with the technology. It would be expected that because access to information is much more easy, that we would have a whole lot more to talk about but that is not what is happening. Humans are becoming more limited in their imagination and this is something we should not allow it to happen.

Expanding Imagination:

Albert Einstein which became the synonymous to genius stated that imagination is bigger than knowledge. All the Technology that exists today is proof that when humans use their imagination and have faith in their Science they can conquer wonders in nature. All of Science today was just a deep imagination on human mind in the past. It is the belief that all natural phenomena can be explained with reason, and that all things can be measured by mankind is what led to the Scientific Advances in the last centuries. If we believe that things are beyond our comprehension and close our minds in believing that imagination can not become true, we are closing doors to progress. There is a thing called faith in Science. All scientists who wrote their thesis papers had faith that their reasoning was right. When shooting arrows in the dark you might hit the target. It is by exploring your imagination that you make the once impossible now possible. Imagination can be good or bad, however. Humanity needs to know how to identify what is just and safe, and separate that from what is evil, dangerous, and unfair. Humans could, for example, benefit from Artificial Intelligence, but there are always dangers associated with Technology. That is why we should always follow a Standard and Rules.

Following a Standard:

Everything needs a structure to exist or survive. The Human Body is composed of a rich and complex system of cells, tissue, organs, glands, veins, and arteries that allows it to live. The Computer is also composed of an elaborate system of several parts and pathways for charges to flow. As charges or electric pulses flow, there is a translation of this current and discharges of Energy into a language that tells the Computer what to do. The Human Civilization also needs a structure that seems to be getting fragile over the years. Everyone must follow a set of rules and behave within well stablished standards that is known as the most right and fair. Education should follow a standard with documents, files, and lessons that every educator uses in their teaching process, and this is passed from generation to generation. Families need a well-established approach and respect for the moral values in a person's life that responsible parents should known, and which guides their nurture and care for their children. Knowledge should pass from generation to generation and this system must be consistent and any changes made to it must be for the better and not for the worse. In the last few decades humanity has gone back in time and not forward with respect to the moral values that we are losing.

World Health:

The world is sick and needs healing. To solve all of the world problems a human alone will not be able to reach success. The lack of Philosophy and good understanding of morality is the leading cause of chaos due to hatred and ignorance. The problem has two sides and neither one is good. One side is called hatred and the other side is called ignorance. Only Philosophy and a good grasp of wisdom at a higher level is capable to bring order in the current social disorder. The Schools need to be a place of education not only of academic subjects but also of social norms on what is ethically correct. People need to be taught how to behave properly. Children and young people should be trained on how to act in diverse life circumstances, on how to praise the national flag, and the power that exists in Democracy where people have a voice in the World Civilization. Events in history are like falling dominos. They began with one person, then a group of people, and later gain popularity, expanding to other groups of people, until eventually spreading all over the world. It is easier to destroy something than to build it. The current world order is in danger of crumbling apart. It is best to build a world order for greater stability in modern times before a possible collapse.

Kingdom of Heaven Control of Space Time

Imagine a very advanced human civilization with such a high level of Scientific understanding that is capable of controlling the Space Time Dimensions around the Solar System. Suppose that this advanced civilization is constantly watching and monitoring human history and when these beings do interfere, it is mostly without our awareness. This invisible civilization in the Solar System makes changes to human history and is responsible for the major world events. These invisible beings are the root cause of many events in world history and they have been observing humans and leading to all major leaps in Technology and Science. This project has been an ongoing process since humans were apelike and it was the work of these beings what guided the biological evolution towards the Intelligent Life on Earth. What if the devil is an extraterrestrial army wanting to destroy humanity as a revenge of something terrible that happened in the past in several Star Systems and Planets in our Galaxy? What if the good beings are currently fighting this demonic extraterrestrial army, and both armies are competing for the human soul on Earth? What if? All of this is just a what if, but what if it is true?

Invisible Civilization:

Suppose that there is an advanced Civilization that has total control of the Space Time Dimensions surrounding the Solar System and nearby Stars. These are Angels that can travel both forward and back in time and they are what protects our Solar System region from dangerous Aliens that wants our destruction. In order to have control of the Space Time in a Galactic Region, these Angels built a Wormhole Web, that allows travel to other Stars and also to particular moments in Earth's History. These Angels would then live in Satellite Cities in Space in orbit of the Sun, and be hidden in many Asteroids and Planets without our knowledge. The constant struggles that humans go through in life is this constant battle between good and evil. It is this chaotic war of the worlds within our world, within our soul. The evil army wants to control humans, while the Angels of Life wants our Freedom. Is Communism a form of control? What exactly guides major historical events and what is the future of humanity? Is there the danger that the forces of good can lose? **Are we in danger? Why are there so many people worried about salvation?** Are we in jeopardy?

Galactic War:

If an invisible civilization has total control of the Space Time Dimensions around and near the Earth, they have total control of all things that happen in our civilization. All of this is just a trip through imagination and nothing like that is probably true but it is always fun to think about possibilities. The amount of Technology we currently have, and how fast we have evolved Scientifically in less than a century is very scary. It only shows that Humans are very intelligent and that we can all things. If the 20th Century was the Age of Technological Progress, the 21th Century will the Age of solving major Social Problems and the developing of even more Technology. Travel to Mars and the Moon should become more and more common and even sending Humans to Space far from the Earth. The construction of Satellite Cities in orbit of Earth, Mars, Venus, Jupiter, and Saturn. Even Satellite cities in orbit of the Sun between the orbits of Mars and Jupiter. The Computer tells me that when humans desire something they are able to reach that goal. Or Galactic War is against our Social Problems, our inequality, our diseases, our enemies since forever. The destiny of an Intelligence Race is reaching Utopia on Earth, and expand our seed throughout the Cosmos.

A World Needing Stability:

The problems that the world suffers are not only in social inequality but also illness of the soul. It is extremely hard to convince a dictator to do what is good, and several people are more likely to never regret anything wrong that they did in the past. The human mind is often very closed and not willing to change the way of thinking, or not wanting to learn how to do what is always fair. If Technology were able to fix all the problems in the world, we would have already done it, but the real problem relies within human beings. The way humans are and what the majority of the people do or think can explain all the chaos seen in the world today. The solution of these problems is not easy. It is like wanting to build a tower but many people destroy with every brick that is placed in the building leading to a fruitless situation. That leads to the inability of us to build that tower since not everyone is willing to help. Same can be said about many nations with a minority trying to help establish a more fair and prosperous government, but the majority of the population refuses to listen, and do not follow the law, and the nation is left fruitless, and in deep poverty. It impossible to convince the majority to do what its best for the nation.

The World and Quantum Physics:

The human tendency to reject change for the better should not prevent us from dreaming of a perfect world in the future which may happen after a very long time. The situation is critical specially in nations where not only people are not willing to change but their culture makes it incredibly hard to allow any progress. We live in a complex reality which can be seen as being rather simple. One trick into solving complicated problems is by looking at them to notice any hints of simple elements within its structure. If for example, you notice several small steps that leads into solving a complicated puzzle, you are noticing that through simplicity you can over time and with patience solve the puzzle. All things in reality whether complicated or not can be understood through reason and faith in reason. The main goal of success is to reach at least some success which over time can lead to a higher success. The universe is like an immense wave and fighting against it is not worth it. You should instead have patience that over time, through small steps can arrive at a great result. Quantum Physics teaches us that at the very small scale all particles exist is a probabilistic manner. There is a probability that everything will end with good results, and that it requires faith in reason through several small steps. You eventually profit from it.

Patterns in the World:

The universe is filled with patterns that exists throughout infinity. The entire cosmos is a fractal geometry and when Scientists are able to measure and define these patterns, they are able to make predictions and explain the meaning of their findings. In Calculus, an Integral is nothing but an Infinite Sum of the Infinitely small that when added together gives the total area or volume of a shape within a Space Time Dimension.

$\sum_{dx \to 0}^{\infty} F(x)dx$ An Infinite Sum of the Height of a Function times the Width dx of an infinitely thin rectangle. The sum of these Infinitesimal rectangles gives you the area under the curve.

$\sum_{dx \to 0}^{\infty} F(x)dx = \int F(x)dx$

Fractal Geometry are shapes that are Infinitely small and others that are infinitely large and they are patterns that flow continuously towards infinity in Space. The universe is a fractal and the cosmic web containing Galaxies may go on and on forever bondless.

Area Under the Curve:

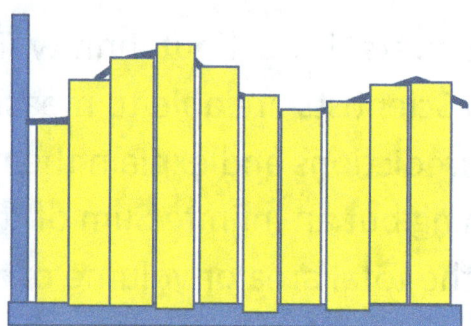

Integrating is the same as finding the sum of the areas of the rectangles with an Infinitely small Width. When approximating the sum towards infinity the area converges towards a value.

There are universal tricks such as the rules governing Integrals that allows calculations to be performed more easily. A person who sees all things believing that these phenomena are beyond human reach would conclude that summing up rectangles would a tedious job for multiple cases in Mathematics. The fact is that the rules of the Integral are derived from known patterns in Algebra that when applied, reduces the amount of work. The Integral makes now not necessary to perform the summation but only to apply known rules from known patterns towards the solution for the area under the curve.

Wave Functions:

All things that are oscillatory in nature can be represent with Sinusoidal Wave or a combination of both Sine and Cosine.

A Sinusoidal is any function which a Sine or Cosine, and Oscillations can be a combination or Superposition of both.

$$\psi(x) = ASin(x) + BCos(x)$$

The graph of the above function assuming that A and B are both equal to 1 is the following:

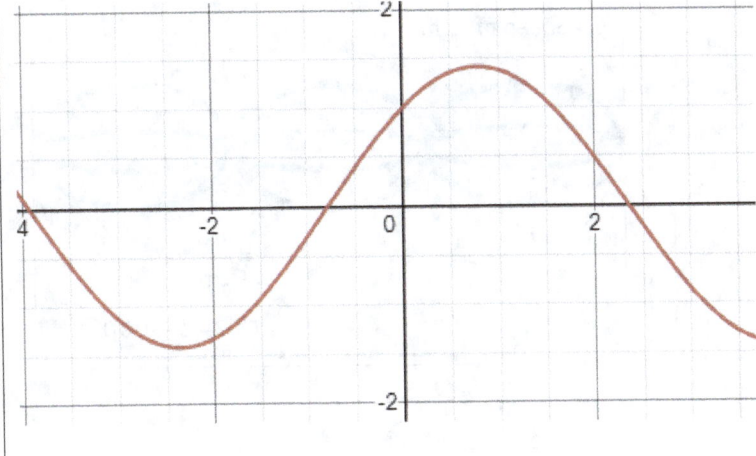

The values of A and B can be adjusted and also the Angular Shift in order to be able to map an infinite number of Wavefunctions. With the advent of String Theory, Scientists are currently looking for a Mathematical Description that will be able to map the Universe's Real Wave Function which is a collection of Waves that comprised the entire Cosmos through Hilbert Space.

Maxwell's Equations

Divergence Theorem:

$$\int (\nabla \cdot F)dV = \oint F \cdot dA$$

The Divergence within a Volume = Field coming from an enclosed Area

$\nabla \cdot \nabla X A = 0$ The Divergence of a Curl is zero since cos(90) = 0

Stoke's Theorem:

$$\int \nabla X F \cdot dS = \oint F \cdot dr$$

The curl through a line path= Dot Product through a closed loop

Divergence Theorem:

$$\nabla \cdot D = \rho \qquad \nabla \cdot B = 0$$

There are no Magnetic Monopoles

$$\oint D \cdot dA = \int \rho dV$$

$$\oint B \cdot dA = 0$$

$$D = \frac{Q}{4\pi d^2}$$

$$D = e_o E + P$$

D = Electric Displacement

B = Magnetic Field

Q = Charge

E = Electric Field

e_o = Dielectric Constant

A = Area   V = Volume

Stoke's Theorem:

$$\nabla X H = J + \frac{\partial D}{\partial t}$$

$$\nabla X E = -\frac{\partial B}{\partial t}$$

$$\oint E \cdot dl = -\int \frac{\partial B}{\partial t} ds$$

$$\oint H \cdot dl = I + \int \frac{\partial D}{\partial t} ds$$

$$\nabla \cdot J = -\frac{\partial p}{\partial t}$$

H and D are Perpendicular to each other. A change in D leads to a change in H. A change in H leads to a change in D. An Electromagnetic Wave.

J = Current Density

H = Magnetic Vector

$H = \frac{B}{u_0} + M$

M = Magnetization

u_0 = Magnetic Permeability

I = Current

Divergence of Current is the negative of the change in charge density over time

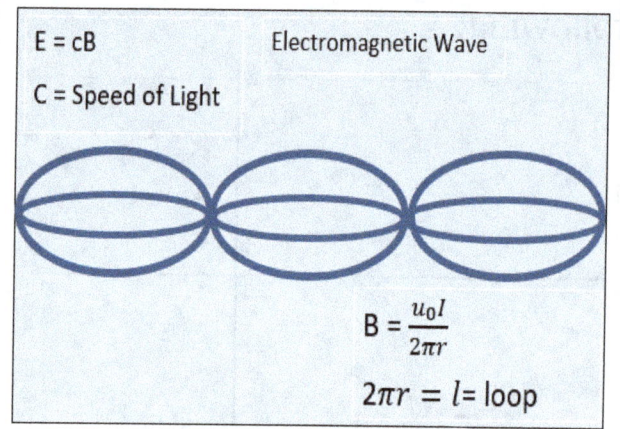

E = cB

C = Speed of Light

Electromagnetic Wave

$$B = \frac{u_0 I}{2\pi r}$$

$$2\pi r = l = \text{loop}$$

Maxwell's Equations Explained:

Light is a basic constituent of Matter. Since all Matter is Energy according to Einstein's famous equation E = MC^2, and since Light is the most basic form of Energy, and since all Particles have Energies, then all Particles are made of Light, and so then is everything a form of Light. Like it says: Let there be Light and everything came into being from nothing in the emptiness of Space. Light has a maximum Speed in the Vacuum which is $3x10^8 m/s$, and it is composed of a Fluctuation of Electric and Magnetic Fields at 90 degrees to each other. When there is a change in the Electric Field, a change in the Magnetic Field then follows since both fields are intertwined. Unlike an Electric Field which diverges from a Positively or Negatively Charged Particle, the Magnetic Fields only Curls, and it is not possible to have Magnetic Monopoles. The amount of Divergence from an Electrically Charged Particle is proportional to its charge. The Curl of the Electric Field generates a change in the Magnetic Field like around a Solenoid. A Curl of the Magnetic Field by changing the Magnetic Flux of a region in Space leads to a Current due to an Electric Field.

Maxwell's Equations is the reason why I believe in the Ether through Space. Only a material which is not entirely empty can be used as a medium for the Electromagnetic Disturbance to propagate from a Radio Station's Antenna to a Radio. **In the same way that you need a coil of wire in order to introduce the Magnet leading to a Current, Space must be made of something in order for Light to flow.** If there is a flow of Electromagnetic Radiation, there is a Disturbance in Space through which this Fluctuation flows, and that Space is the Ether. Only something can be disturbed, which means that Space is not Empty. Since the Ether also bends due to Gravity, this Space is used not only for the Propagation of Gravity but also for the propagation of the Electromagnetic Radiation. Understanding Space will lead to a more complete understanding of the Four Forces of the Universe including making wise steps into the unification of all Four Forces in the Theory of Everything. The Ether can't be detected since it is Space itself and it goes right through Matter without generating an Ether Wind. When two Neutron Stars collide they generate a Gravitational Disturbance in Space which propagates by bending the Ether in a Gravitational Wave sort of like an Ether Wind. **Maybe to feel the Ether Wind the collision of Black Holes or Neutron Stars are required, and anything as small as Earth moving through Space is not enough for there to be the detection of the Ether Wind.**

Space is Made of Something:

In the same way that Gravity which is the weakest of all Forces of Nature requires an object the size of a Planet to generate a noticeable Force and Acceleration, the Ether Wind requires a major event that can generate powerful Gravitational Waves from major Interstellar and Intergalactic Explosions. Anything smaller than that does not generate an Ether Wind that can be felt. Trying to measure the Ether Wind by using Earth's motion around the Sun is like wanting to Measure the Force of Gravity due to an ant on the ground, just too little. The Ether goes right through Matter and the compression of the Ether leads to Matter and the stretch of the Ether leads to what we would call empty Space. Filled Space is made of the Ether compressed leading to Waves in Strings and the present Energy and Matter. Empty Space is made of the Ether that is stretched and no Waves on Strings, thus what we would call nothing.

Gravitational Waves is the Ether Wind

What is the Ether?

The Ether has probably already been discovered and Scientists are not aware of it. The Higgs Bosons are Particles that fill Space within a Higgs Field. The Higgs Boson is what gives mass to Particles. **When Higgs Bosons Cluster, they give mass to Particles, and that is very similar to my idea that when the Ether Compresses it gives shape to Matter and Energy as we know it**. When the Ether stretches there is no mass present. When the Higgs Bosons move apart in a region of Space, it signifies that the region in Space is probably empty. I imagine the Higgs Bosons and the Higgs Field to be the Ether. **They generate Mass which eventually generates Gravity so that means that there is no need of a Graviton. A Higgs Boson is all you need to explain Gravity**. Higgs Bosons are Bosons just like Photons of Light are Bosons, and possibly the Higgs Bosons are also made of Light, and since Light propagates in Space it uses the Higgs Bosons to propagate. The Higgs Boson is a good candidate to explain all Four Forces of nature since all Four Forces use the Higgs Field, the Ether as a means of Propagation.

Center of Gravity:

The Center of Gravity in a Solar System is the center that pulls all objects around it including the Sun which is not the Center of Gravity. The Sun Wobbles around the Solar System's Center of Gravity and all Planets orbit this Center of Gravity outside the Sun. The Sun's Wobble is what allows Scientists to map and estimate the number of Planets that orbit it. Planets of different masses and different distances from the Sun will lead to different Wave Patterns in that Sun's Wobble around the Center of Gravity. In an earlier example I explained that massive objects moving in a circle at high speeds close to the Speed of Light will generate Gravity in the Center of the Circle that it rotates which happens to be the Center of Gravity of the System.

It has already been proven that all Planets, Stars, Galaxies, and Massive Objects spin or rotate around the Center of Gravity like as if the Center of Gravity is pulling them by keeping them in Orbit. If a mass rotates really fast near the Speed of Light, it can generate Gravity at the Center of the rotation.

Center of Gravity

Time:

The Flow of Time occur when Time at the Speed of Light passes an object.

The closest to the Speed of Light that an object moves, the Slower is Time able to pass by the object.

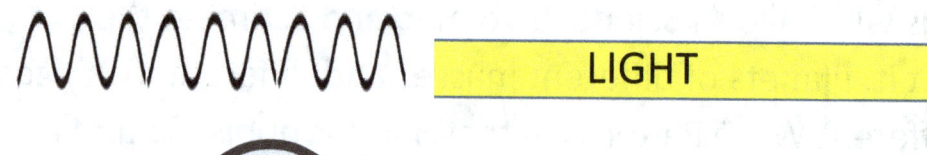

LIGHT

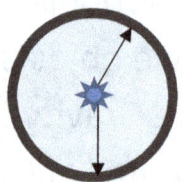

Time

Time flows at the Speed of Light. When something moves at the Speed of Light, Time is unable to catch that object, and Time for that Object stops. For a Photon of Light there is no Time and neither is there Space.

The Four Forces of Nature:

Strong Force:

Keeping Nucleus of Atoms bond together.

Weak Force:

Breaks up the Nucleus of Atoms causing Decay and Radiation.

Electromagnetic:

Photons of Light, the Electric and Magnetic Field between Charges.

Gravity:

Attraction between Masses.

Particles of the Forces of Nature:

Strong: Protons and Neutrons are made of Quarks. Gluons between Quarks bond Protons and Neutrons together in the Nucleus of Atoms.

Weak: The W and Z Gauge Bosons leads to the decay of Atomic Nuclei emitting radiation and leading to aging.

Electromagnetic: Photons of Light emitted by Particles generate an Electric Field causing attraction and repulsion between Particles. The Electric and Magnetic Fields are intertwined in the Electromagnetic Radiation as explained in Maxwell's Equations.

These three Forces of Nature are mediated by Bosons, Photons of Light that propagate using the Strings in the Ether of Space. The Direction of the Spin of Particles generate a disturbance in the Waves of the Strings in the Ether Space leading to their Charges. Neutral Charges Spin enough to Interact with a Magnetic Field but are still left with no Charge.

Gravitational: The Higgs Bosons cluster in Matter, leading to waves in the Ether of Space. These waves in Matter pull other Waves in other Matter leading to attraction between Masses.

Oscillations in a Spring:

Suppose there is the Oscillation of a Spring Mass System:

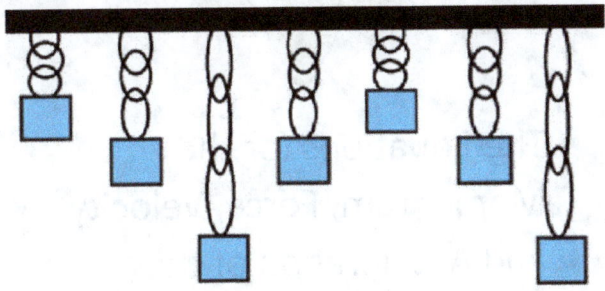

The Oscillation can be represented with a Cosine Wave with Angular Frequency equal

to: $\omega = \sqrt{\dfrac{k}{m}}$

Where K is the Spring Constant and m is the mass of the Hanging Mass.

As the Mass Oscillates it gains and loses Momentum. The Momentum is Maximum where Kinetic Energy is Maximum at the Equilibrium Positions. The Momentum is Minimum where the Potential Energy is Maximum and the Force the Greatest.

Momentum $P_y = mv_y = \text{m}\dfrac{dy}{dt}$

$$\frac{d^2y}{dt^2} = \frac{d}{dt}\left(\frac{dx}{dt}\right) = \frac{d}{dt}\left(\frac{P_y}{m}\right) = \frac{ma_y}{m} = a_y = -\omega_0^2 y$$

$$\omega_0{}^2 = \frac{k}{m}$$

The Force of the Spring on the Mass is:

$$\text{Force} = ma_y = m\frac{d^2y}{dt^2} = -ky$$

Where y is the Displacement from Equilibrium.

$$\frac{d^2y}{dt^2} = -\frac{k}{m}y = -\omega_0{}^2y$$

And

$$\frac{dy}{dt} = \frac{P_y}{m} = \frac{mv_y}{m} = v_y$$

And

$$\frac{dp_y}{dt} = -m\omega_0{}^2y = m\frac{d^2y}{dt^2} = ma_y = \text{Force}$$

The Equations for the Momentum, Force, Velocity, and Acceleration of the Oscillating Pendulum is shown on the Calculations on the left.

ZIRVINA

The Hamiltonian of the Mechanical System is:

$$H(y, P_y) = U(y) + K(P_y) = \frac{1}{2}m\omega_0^2 y^2 + \frac{P_y^2}{2m} = \textbf{Total Energy}$$

$U(y) = $ Potential Energy

$K(P_y) = $ Kinetic Energy

Which leads to:

$$\frac{\partial}{\partial y}H(y, P_y) = \frac{\partial}{\partial y}(\frac{1}{2}m\omega_0^2 y^2 + \frac{P_y^2}{2m}) = m\omega_0^2 y = ma_y = \textbf{Force}$$

And

$$\frac{P_y}{m} = \frac{\partial}{\partial P_y}(\frac{1}{2}m\omega_0^2 y^2 + \frac{P_y^2}{2m}) = \frac{mv_y}{m} = v_y = \textbf{Velocity}$$

Partial Derivatives on the Hamiltonian can be used to solve for the Forces, and Velocities of the Oscillating Spring-Mass System.

Euler Lagrange Equation:

$$\frac{d}{dt}\left(\frac{\partial L}{\partial v}\right) - \frac{\partial L}{\partial y} = 0$$ and L = Kinetic Energy – Potential Energy

With $L = \frac{1}{2}mv^2 - \frac{1}{2}ky^2$ For the Spring System

$$\frac{\partial L}{\partial v} = \frac{\partial}{\partial v}\left(\frac{1}{2}m\omega_0{}^2y^2 + \frac{P_y{}^2}{2m}\right) = \frac{\partial}{\partial v}\left(\frac{1}{2}m\omega_0{}^2y^2 + \frac{(mv)^2}{2m}\right)$$

$$= \frac{\partial}{\partial v}\left(\frac{1}{2}m\omega_0{}^2y^2 + \frac{m^2v^2}{2m}\right) = mv$$

$$\frac{d}{dt}(mv) = ma = Force$$

And

$$-\frac{\partial L}{\partial y} = -\frac{\partial}{\partial y}\left(\frac{1}{2}m\omega_0{}^2y^2 + \frac{P_y{}^2}{2m}\right) = -m\omega_0{}^2y = -ky = Force$$

Force – Force = 0

Force = - ky   thus proving once that $\frac{d}{dt}\left(\frac{\partial L}{\partial v}\right) - \frac{\partial L}{\partial y} = 0$

Entangled Equations and Space:

Mathematics is the study of numbers and patterns in the Universe. Physics is applied Mathematics where it is used to make sense of Physical Phenomena. From the First Law of Thermodynamics the Hamiltonian is constructed as the Total Mechanical Energy of a System. Energy can't be created nor destroyed only transformed and the Hamiltonian is composed of Kinetic and Potential Energy added together. In this case, Friction and Energy losses during Oscillations are neglected and in a completely Closed System, the Hamiltonian becomes constant thus preserving the Energy. No true closed system exists, unless the Universe as a whole is a Closed System, but anything within the Universe loses Energy thus increasing the Entropy or Disorder of the Universe. Energy in a Closed System then Oscillates between Potential and Kinetic in complete cycles. The Partial Derivatives of the Hamiltonian can be used to find the Force leading to the Oscillation and the Velocity of the Mass at a particular points in the cycle. Both the Force and Velocity changes during the Oscillation. Maximum Force happens at Maximum Potential Energy, and Maximum Velocity where the Force is zero at equilibrium position. The Euler-Lagrangian Equation reveals a pattern where it relates Partial Derivates of the Lagrangian to the Force and Momentum of a System. The fact that Mathematics is able to relate one Equation to the other assuming a Closed System, or a Perpetual Motion Machine which is not realistic, but it is useful into understanding through this Ideal Case, patterns that can help us see how the Equations are really entangled to Variables and yet other Equations. Despite the real world not being anywhere an Ideal Case, and being very chaotic, the understanding of Ideal Cases can still help Scientists have an idea or at least a Mathematical Approximation to the Laws that governs the Cosmos. **There is no Perpetual Motion Machine and the Hamiltonian assumes just that as an Ideal Case.** Mathematics in Science is still a very good approximation to the chaotic motion of Particles in the Universe. We can always assume that if our measurements are off from the results in the experiment, it is and indication of the rise in Entropy and that is also one the Laws of Thermodynamics.

 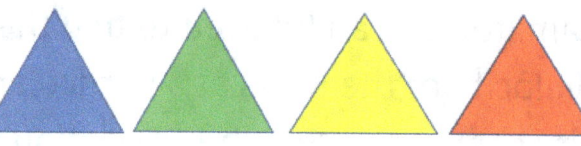 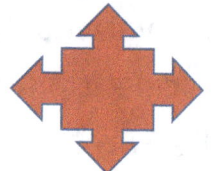

Laws of Thermodynamics:

Zeroth Law:

If a material A is in Thermal Equilibrium with material B and B is also with C then A is also in Thermal Equilibrium with material C.

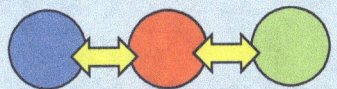

First Law:

Energy cannot be created nor destroyed only transformed.

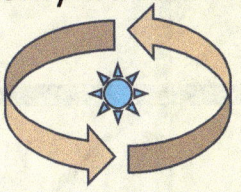

Second Law:

Entropy in an Isolated System always increases. There is no Perpetual Motion Machine. Energy is always lost due to Friction, Heat, Resistance, and so forth...

Third Law:

Temperature at Zero Entropy is Absolute Zero at 0 Kelvins or -273 Degrees Celsius.

Phase and Real Space:

When graphing the Oscillation of the Spring Mass System we get the following:

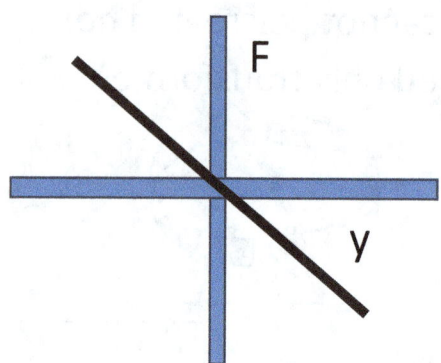

Where F is the Restoring Force on the Spring and y is Displacement from Equilibrium.

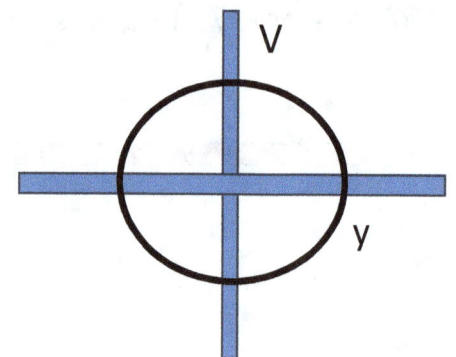

Where V is the Velocity and y is the Displacement from Equilibrium assuming no Energy Losses. At Maximum Displacement v = 0, and Force is Maximum. At zero Displacement V = Max and Force is Zero.

At zero Displacement Kinetic Energy is Max, and at Maximum Displacement Kinetic Energy is zero.

A zero Displacement Potential Energy is Zero, and at Maximum Displacement Potential Energy is Max.

Real Oscillations are Damped and Energy is lost to the environment according to the Second Law of Thermodynamics. With Damped Oscillations the Phase Space Graph will swirl towards the center reaching the Zero Point, and all other graphs will converge towards no Energy, and no more Oscillations.

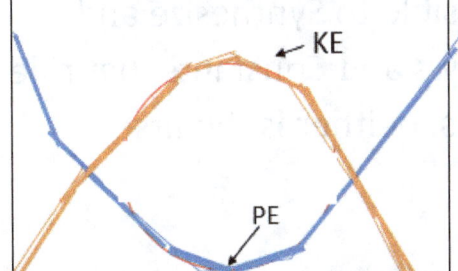

Energy of Spring Mass System

KE

PE

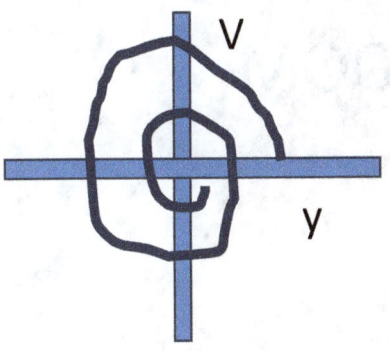

V

y

In Phase Space no lines on the graph should cross.

Short Conclusion:

In this book I have introduced a variety of topics that revolve around the idea of Oscillations and a desire to have a summary of the essence behind all natural Phenomena and the use of Mathematics to understand the discreteness within the chaotic motion of Particles and Matter in the Universe. It is the faith that all knowledge can be within human grasp, and that it is possible to Synthesize and summarize what we know as an approximation to the Laws and Constants that rule the cosmos. This is not the last book about topics like this, neither is the first.

<u>About the Author</u>

I, Diogo Franklin de Souza, was born in the city of Rio de Janeiro, Brazil in August 20, 1986. I moved to Dallas, Texas when I was 11years old. I write stories since I was 9 years old. My books tend to contain short summaries of the most important things I find about life, morality, philosophy, and science. Like I say, everything is part of a whole system, and this is also for everything I do and write. I always wanted to have all the most important knowledge in only a few short books. That is why I write, and that is my inspiration for short summaries. I hope this book brings some inspiration also for the readers, because that really is the purpose of my work. Read it and take from it, pieces of gold for you that can be useful in your life. Enjoy….

JJJJ